BRITISH WEATHER AND
THE CLIMATE OF ENLIGHTENMENT

British Weather and the Climate of Enlightenment

{ JAN GOLINSKI }

The University of Chicago Press · Chicago and London

The University of Chicago Press, Chicago 60637
The University of Chicago Press, Ltd., London
© 2007 by The University of Chicago
All rights reserved. Published 2007.
Paperback edition 2011
Printed in the United States of America

20 19 18 17 16 15 14 13 12 11   2 3 4 5 6

ISBN-13: 978-0-226-30205-8 (cloth)
ISBN-13: 978-0-226-30203-4 (paper)
ISBN-10: 0-226-30205-9 (cloth)
ISBN-10: 0-226-30203-2 (paper)

Library of Congress Cataloging-in-Publication Data
Golinski, Jan.
British weather and the climate of enlightenment / Jan Golinski.
    p. cm.
Includes bibliographical references and index.
ISBN-13: 978-0-226-30205-8 (cloth : alk. paper)
ISBN-10: 0-226-30205-9 (cloth : alk. paper)
1. Meteorology—Great Britain—History. 2. Great Britain—Climate.
    3. Weather. 4. Climatology. I. Title.
QC989.G69G66 2007
551.50941′09033—dc22
2006031722

Pray what was that man's name,—for I write in such a hurry, I have no time to recollect, or look for it,—who first made the observation, "That there was great inconstancy in our air and climate?" Whoever he was, 'twas a just and good observation in him.—But the corollary drawn from it, namely, "That it is this which has furnished us with such a variety of odd and whimsical characters;"—that was not his;—it was found out by another man, at least a century and a half after him: Then again, . . . that this strange irregularity in our climate, producing so strange an irregularity in our characters,—doth thereby, in some sort, make us amends, by giving us somewhat to make us merry with when the weather will not suffer us to go out of doors,—that observation is my own;—and was struck out by me this very rainy day, March 26, 1759, and betwixt the hours of nine and ten in the morning.

LAURENCE STERNE · *The Life and Opinions of Tristram Shandy*

{ CONTENTS }

$\{$ ILLUSTRATIONS $\}$

THE WEATHER IS ALWAYS WITH US. Even though many of us in modern society spend most of our time indoors, we cannot escape it. Remarks about what it is doing or about to do smooth our everyday social interactions. Reports, observations, and predictions punctuate our daily routines. Extreme weather fascinates us with its uncontainable violence, now brought near by television even when it occurs on the other side of the world. Modern technology has shielded us from some of these dangers; at least we can be forewarned of the approach of thunderstorms, hurricanes, floods, and tornadoes. But this increased knowledge has not necessarily brought increased security from the weather's effects. In certain respects, modern society has placed itself more in the way of harm by the weather, which continues to force itself upon our notice and constantly threatens to disrupt our comfortable lives.

Quite a lot of attention has recently been devoted to the cultural signifi-cance of our weather worries.[1] In part, the preoccupation expresses prevail-ing concerns about modern life itself, which is thought to have exposed us to new hazards by trespassing upon the natural environment. As the French philosopher Michel Serres has put it, "Today our expertise and our worries turn toward the weather, because our industrious know-how is act-ing, perhaps catastrophically, on global nature."[2] Recent disasters, such as Hurricane Katrina, along with more persistent anxieties about the in-fluence of human activity on global climate change, have reinforced the point. Journalists and academics have echoed Serres's suggestion that the weather's intrusions — aside from their immediate, sometimes catastrophic impact on those directly affected — also raise the fundamental issue of humanity's relationship to nature as a whole.[3] The question is raised of the degree to which human beings have mastered the natural world and how we understand ourselves in relation to it. The weather confronts us with a challenge to the intellectual reach of modern science, to our technological capabilities, and, more basically, to our sense of ourselves as at home (or not) in our physical environment.

In this book, I take these concerns as an invitation to historical inquiry.

It seems worth asking how this situation came about. I propose that we look to the eighteenth-century Enlightenment as the era when fundamental characteristics of modernity and its symptomatic attitudes to the natural world were forged. I want to see how understandings and experiences of the weather figured in the process of enlightenment and modernization. I am focusing this inquiry on Britain, with some references to its overseas colonies. This is partly for pragmatic reasons of my own knowledge of the primary materials and archival resources. I am fully aware that aspects of Britain's history in the eighteenth century, of its involvement in the Enlightenment, and indeed of its weather are unique. In fact, the specific features of the British cultural and social setting will be part of the story I shall tell. The focus on a single nation, though it yields in some respects a partial picture, enables me to give a richer account of the cultural meanings ascribed to the weather in this period.

The argument of the book has three strands. First, while the idea of climate is an ancient one, it was reconceived in the eighteenth century through systematic study that attempted to normalize the weather, to reduce it to some kind of regularity. This development began in Britain in the late seventeenth and early eighteenth centuries. Inspired by leading physicians and natural philosophers, observers for the first time devoted attention to recording the weather on a daily basis. Meteorological instruments, including thermometers and barometers, were introduced and widely circulated. The weather came to be attended to as an item of "news": a topic of discussion in the new print media and a subject of public conversation. Through renewed attention to their weather, the British forged a new understanding of their climate, which became a component of the emerging sense of national identity. The British climate was regarded as a providential asset of the nation, a divine gift to the population's health and prosperity.

Second, discourse about the national climate spoke to the British people's sense of living at a time of significant historical change. Thinking and speaking about their weather, the British were also alluding to the processes that were changing their society, including those that we might say constituted the experience of enlightenment. Thus, in certain respects, attitudes to the weather and climate in the eighteenth century appear unmistakably modern. Enlightened investigators detached the weather from expectations of impending apocalypse or fears of divine punishments. They subjected it to routine, meticulously recording conditions on every day and measuring them with instruments. They insisted that the climate reflected the regular actions of physical laws that were manifestations of God's providential benevolence. The climate assumed a fundamental role in Enlightenment thought about society and history. It was recognized as a material influence on human health and welfare that significantly affected the development of the world's peoples. Eighteenth-century thinkers speculated about how cultural and material causes interacted in social progress; they bequeathed to subsequent social theory an important legacy of reflection on the role of the physical environment in human history.

On the other hand, however—and this is the third strand of the argument—thoughtful people were constantly made aware in this connection of the ways in which the process of enlightenment fell short of its most ambitious goals. The weather could never be entirely reduced to regularity; its anomalies and extremes continued to spring surprises. And when they did so, they evoked attitudes among the masses that enlightened intellectuals tended to deplore as primitive superstitions. Even British society showed a distinctly unenlightened face when confronted with violent or unusual weather. In other respects, too, the climatic influences on human life showed how much it was determined by natural forces that would not submit to the powers of reason. Thus, reflecting on climate and the weather, enlightened intellectuals recognized the constraints on rationality, the mind's dependence on the physical body, the limited accomplishments of cultural and social reform, and generally the interdependence—the continuing inextricability—of nature and culture.

It is this which makes the eighteenth century an appropriate mirror of our own age and allows a study of the British experience to yield up its more general implications. To this day, the weather remains unpredictable to a significant degree. Its occasional manifestations of extreme violence appear as reminders of the powers of nature that evade human control. It continues to affect our health and emotional state in ways we cannot entirely understand and to unsettle our confidence in the power of reason. Our society's vulnerability to meteorological crises and catastrophes shows that scientific rationality has never completely mastered the natural environment. Hence, the weather has come to bear the burden of some of our most profound concerns about modern society, its past transgressions, and its future prospects. The basic claim of this book is that contemporary attitudes to the weather—our unease about its dangers, our sense that these reflect something profound about our relationship with nature—are not as new as we might imagine. Eighteenth-century intellectuals already realized that the weather's unpredictability demonstrated the limits of human understanding and control of nature. Even while they worried that social change was leading them away from a natural mode of life, they were obliged to recognize the persistent power of the natural forces that underpinned human existence. Anxieties about weather and climate expressed both a sense of the incompleteness of the process of enlightenment and qualms about the consequences of its successes. Our current weather worries look quite familiar in this context. Looking back to the eighteenth century from this standpoint, one might adopt Immanuel Kant's famous judgment and say that we still live in an age of enlightenment but not yet in an enlightened age.[4]

It is a pleasure to acknowledge the many debts I have accumulated over the years it has taken me to complete this book. My home institution, the University of New Hampshire, has substantially assisted in several ways. I was appointed Director of the UNH London Program for the year 2000–2001, which allowed me a period of research in London libraries and archives. I received a fellowship from the Faculty Scholars Fund for the spring semester of 2004 and a sabbatical leave in the fall of that year. I am also grateful to the James Fund of the Annual Alumni Gifts Fund of the College of Liberal Arts for a subvention to support publication of the book, and to the Rutman Fund of the Department of History for a grant to help with the cost of illustrations. I particularly appreciate the work of the Dean of Liberal Arts, Marilyn Hoskin, and the Chair of the Department of History, Janet Polasky, in helping secure these grants. I also wish to thank my colleagues in the Department of History and in the Humanities Program for their encouraging and probing responses to presentations on the project.

The Centre for Research in the Arts, Social Sciences and Humanities of the University of Cambridge awarded me a visiting fellowship for the Michaelmas Term, 2004. I particularly appreciate the assistance of the staff of the Centre in easing my relocation to Cambridge, the collegiality of my co-fellows for that term—especially my housemate Jan Birksted—and the generous hospitality of the director, Ludmilla Jordanova, and the deputy director, John Morrill, who together created a model environment for interdisciplinary inquiry.

The following institutions hosted presentations on aspects of the project: Massachusetts Institute of Technology; Cornell University; Max Planck Institute for the History of Science, Berlin; University College, London; University of Cambridge; Freie Universität, Berlin; Gonville and Caius College, Cambridge; University of California, Berkeley; University of Washington; University of Pennsylvania; McGill University; and Harvard University. I thank them all, and I thank the audiences who attended the talks for the stimulus of their comments and questions.

I am grateful to the staff of the following institutions for their assistance with my research: Bodleian Library; British Library; Cambridge University Library; Dimond Library, University of New Hampshire; Hereford and Worcestershire Record Office; Houghton Library, Harvard University; Lewis Walpole Library, Yale University; Royal Society, London; Wellcome Library, London; West Sussex Record Office; and Whipple Library, Cambridge. I especially appreciate the hospitality, extending well beyond normal expectations, of Janet Pennington, the Archivist at Lancing College, West Sussex, and of Michael Wood and Ian MacGregor at the National Meteorological Archive, Bracknell, Berkshire.

At the University of Chicago Press, I benefited from the friendship and encouragement of the late Susan Abrams at an early stage of this project. Catherine Rice has proven no less a friend and a skilled and tactful editor in seeing the book through. I also appreciate the advice of two anonymous referees appointed by the press.

I should acknowledge that earlier versions of some of the following chapters have appeared in print: chapter 1 in " 'Exquisite Atmography': Theories of the World and Experiences of the Weather in a Diary of 1703," *British Journal for the History of Science* 34 (2001): 149–71; parts of chapters 2 and 3 in "Time, Talk, and the Weather in Eighteenth-Century Britain," in *Weather, Climate, Culture*, edited by Sarah Strauss and Benjamin S. Orlove, 17–38 (Oxford: Berg Publishers, 2003); and part of chapter 4 in "Barometers of Change: Meteorological Instruments as Machines of Enlightenment," in *The Sciences in Enlightened Europe*, edited by William Clark, Jan Golinski, and Simon Schaffer, 69–93 (Chicago: University of Chicago Press, 1999).

Last but by no means least, I offer heartfelt thanks to the following friends and colleagues for their advice, encouragement, and general support: Ken Alder, Katharine Anderson, Hasok Chang, Catherine Crawford, James Delbourgo, Nick Dew, Brian Dolan, Patricia Fara, Jim Fleming, Richard Hamblyn, Rebecca Herzig, Vladimir Janković, Adrian Johns, Larry Klein, Dorothy Porter, Jessica Riskin, Lissa Roberts, Andrea Rusnock, Simon Schaffer, Anne Secord, Jim Secord, Steven Shapin, Liba Taub, Mary Terrall, Jenny Uglow, Nick Webb, Simon Werrett, and Alison Winter.

Early in the twenty-first century, eighteenth-century studies suffered the loss of two wonderful scholars, Michael DePorte and Roy Porter. Both were generous with their enthusiasm for this project, though sadly neither lived to see the final result. In dedicating the book to their memory, I want to record how the example of their work and friendship continues to inspire those who knew them.

Weather and Enlightenment

The weather is the primary sign of the inextricability of culture and nature. JONATHAN BATE · *"Living with the Weather"*

STIRRING THE ICE IN HIS INKPOT, his fingers numb with cold, a writer living in Worcestershire in the west of England in January 1703 began to compile a diary of the weather. His daily journal, with its vivid and meticulous descriptions of atmospheric events, has survived down the centuries, though it was almost entirely unread and the name of its author was long forgotten.[1] As he wrote, the diarist tried to answer questions that were to preoccupy intellectuals in the century then dawning and into our own times: How are atmospheric phenomena to be explained? Does the weather exhibit regular patterns over the long run, or will it always remain unpredictable? How frequently do violent storms occur, or extremes of heat and cold? Are popular sayings and beliefs about the weather worth attention, or should they be dismissed as simple superstition? How does the weather affect people's health or mood? The diarist could not have known, when he started his journal in January, that November 1703 would bring an excessively violent storm, the most damaging ever recorded in the British Isles. Toward the end of the year, as he surveyed the storm's damage to his own property and read accounts of death and destruction from elsewhere, the writer recorded his melancholy reflections on human weakness in the face of nature's powers.

Eighty years later, another dramatic weather event seized the attention of the British population and reminded them of their vulnerability. During June 1783 a gloomy haze settled in the atmosphere, darkening the sun for weeks. The air became increasingly sultry and oppressive. Undisturbed by wind or rain, the haze made breathing difficult, rotted foodstuffs, and slowed the growth of crops. Many observers recorded the phenomenon, though no one understood the connection with what was later found to have been the cause: a volcanic eruption in Iceland. In Hampshire, the vicar of Selborne, Gilbert White, wrote that the event awakened apocalyptic fears among the populace. At a school near Oxford, a Quaker boy, Luke Howard, was inspired by the anomalous season to a lifelong interest in the weather. Howard later settled in Essex, became a passionate observer of the atmosphere, and invented new ways to record and analyze it. He is best known for developing the nomenclature still used today to classify clouds.[2] He also understood that meteorology intersects with sociology, that finding out about the weather involves listening to what people say about it. In his writings, he often discussed the ways people talked about the weather and how they tried to make sense of it.

These two observers, the Worcestershire diarist and the Quaker meteorologist, were among many British men and women who pioneered the study of weather in the age of enlightenment. As their country took its first steps toward modernization, with burgeoning commerce and the beginnings of revolutionary growth in agriculture and industry, new ideas about the weather came to the fore, assuming a place in the beliefs of a people increasingly confident of their destiny as a civilized nation. As they came to think of themselves as an enlightened people, the British developed a sense of what their homeland owed to its weather. As Howard remarked, "Habit completely reconciles the Englishman to a sky, which rarely glows for a week together with the full sun, and which *drips*, more or less on half the days of the year." Far from being a handicap, however, the damp and chilly air was invigorating, and "incessant changes" of weather were mentally stimulating.[3] The changeableness of atmospheric conditions on the island was generally thought to be a positive influence on the spirit of the people. British writers decided they would rather not have tropical sunshine or heat, which brought with them lassitude, immorality, and disease. A mutable but temperate climate—notwithstanding occasional anomalies—was thought to be good for the country's bustling commercial life and its population's health.

Many of the attitudes to the weather common in Britain today were established during this period. The climatic peculiarities of the island became embedded in the national culture. Sitting as it does between the westerly air currents of the North Atlantic and the more stable atmospheric patterns of the European continent, Britain experiences rapid changes in weather conditions within fairly narrow limits of temperature. It rains a lot, but intermittently and usually not very heavily. Prolonged periods of extreme heat or cold are rare. During the eighteenth century, commentators first noted how these conditions provided material for conversational remarks to acquaintances or strangers. Samuel Johnson protested the pointlessness of such exchanges, but the custom persists. Johnson also used to say that a person of sufficient mental discipline should be able to ignore the weather. His attitude still seems to prevail among those British men who refuse to wear raincoats or carry umbrellas, though foreign tourists understand that these are very prudent accessories in the prevailing conditions. When the British talk about their weather, they draw upon recollections that are tinged by nostalgia and often quite inaccurate. They remember extraordinary events, like the "hurricane" of October 1987. Sometimes they exaggerate seasonal extremes, as with memories of white Christmases that have only rarely occurred. Often, extreme and violent conditions are thought to be increasing and are taken as signs that the climate is changing.[4] Serious storms and floods are believed to be becoming more frequent. Warmer summers, which one might imagine would be welcome, are said to bring the threat of invasion by foreign insects and diseases. Sometimes the British worry that their weather is becoming "more Continental," perhaps a reflection of their hesitancy about involvement with the European Union. These concerns are expressions of uncertainties about the future; they invoke a common notion of the past and the good old British weather, which is feared to be passing away.

In this book, we look back to the age when the British first began to formulate ideas about their national climate based on accumulated records of the weather. Somewhat paradoxically, we shall also find that people then were already saying that the climate was changing. Modernization brought with it the sense of a break from a traditional or natural way of life. One consequence of this was a belief in some quarters that the weather was being altered, for example by new agricultural practices, urban growth, or deforestation. In other ways, too, beliefs about the atmosphere reflected the British people's experiences of change in this era. Spectacular or threatening events, like the storm of 1703 or the summer haze of 1783, aroused anxieties about the cultural changes we associate with the Enlightenment. Unusual atmospheric phenomena—including storms, auroras, peculiar cloud formations, and other heavenly wonders—had traditionally been interpreted as portents of dramatic political events or admonishments of a punitive God. When such things occurred in the eighteenth century, people wondered whether the progress of reason and science had really vanquished such old superstitions.

While dramatic events of this kind seized the attention of the people, the eighteenth century also saw the weather become a topic of more continuous public interest. This followed the beginnings of systematic study of the atmosphere in the late seventeenth century. Like the 1703 diarist, hundreds of individuals began to compile daily weather journals. They downplayed extraordinary episodes in favor of quotidian regularity; they built up comprehensive annals of the weather in their own locality in an attempt to discern long-term patterns. Some of the diarists were medical practitioners searching for connections between the condition of the atmosphere and the occurrence of diseases. Many of them used instruments, especially barometers and thermometers, which were becoming widely available in the growing market for consumer goods. British researchers led the world in this kind of regular recording, and the efforts of these individuals informed public discourse about the national climate. It was generally seen as unpredictable in the short term but moderate overall, capable of sometimes quite rapid changes but rarely veering to extremes. It almost never inflicted excessive heat or cold and seldom generated violent storms. On the other hand, it continuously yielded a good quantity of fertilizing, healthy rain. British observers came to identify their climate with the patterns that prolonged record-keeping had revealed. They concluded that they had been blessed by a benevolent providence with a climate well adapted to sustain the nation's prosperity and well-being.

The records of British observers contributed to a broader investigation of the influence of climate on human life that gathered pace during the eighteenth century. Ancient writers on medicine and geography had discussed how the physical environment of a particular place bore upon its inhabitants' character and health. In the early-modern period, this inquiry was revived and extended as Europeans came to know more of the world's peoples and faced the problem of explaining their diversity.[5] In the eighteenth century, the discourse of climate became entwined with theories of the development of civilization. Climate featured prominently in Enlightenment accounts of national character and historical progress that related specific customs and institutions to the environmental conditions in which people lived. For the British, the question of the influence of climate on civilization was an urgent one and not merely of academic interest. As their nation began to assert itself as a global power, climatic conditions in Africa, India, and the Americas became of pressing concern, since they potentially determined the fate of conquest and settlement. In the parts of the world where the British ventured, the methods of the weather diarists were used to record local conditions. Colonists in North America and elsewhere used instruments and made measurements to compare weather conditions with those in the homeland. Geographical and medical writers also claimed that colonization itself was affecting the climate. Settlement in North America, for example, was believed to be changing conditions for the better, moderating the extremes of winter cold and summer heat. Settlers believed that by improving the land — clearing the forests and draining the swamps — they were also civilizing the atmosphere, compressing into just a few decades developments that in Europe had taken centuries. They expressed the optimism of the Enlightenment in their conviction that the American climate had been tamed by human enterprise and reason.

In other respects, however, eighteenth-century studies of the weather did not encourage expectations of progress. Despite the best efforts of the weather diarists, laws of the atmosphere remained elusive, and extreme and unusual conditions continued to occur unforeseen. Enlightened intellectuals acknowledged that strange weather was viewed with apprehension by the general populace. Throughout the period, the traditional lore of "weather-wising" — invoking proverbs, sayings, and natural signs to foretell the weather — remained in circulation among all sections of the population, unrivaled by any more "scientific" means of forecasting. Medical writers recognized that the qualities of the air affected people at a level beneath rational awareness, modifying their moods and states of health in ways they could not entirely understand. Even the meteorological instruments that were sometimes hailed as tokens of enlightenment were also thought to be objects of superstitious awe. Thus, investigations of the weather and climate testified in some respects to the advance of science and reason, in others to the continuing hold of tradition and the passions. Studies of the air and its effects echoed both the successes and the failures of enlightenment and thus its uneven progress in eighteenth-century society. It is this which makes attitudes toward the weather an index of the often conflicting tendencies of cultural change.

These ambiguities form a major theme in the chapters that follow. They have not often been attended to by scholars. Records of the weather have been studied by historical meteorologists and climatologists, who are interested in what conditions were like in the past, but not specifically in how they were understood by those who experienced them.[6] The cultural dimension of the subject has been left largely to historians of science, who have usually approached it in terms of the development of the science of meteorology. Vladimir Janković's excellent book *Reading the Skies* (2000) has shown how English observers in the eighteenth century sustained the classical preoccupation with the phenomena known as "meteors." They published numerous reports of aerial anomalies—such as auroras, thunderstorms, and whirlwinds—that had long been regarded as characteristic of the middle region between the earth and the heavens. Janković sees the continuing interest in peculiar weather phenomena as evidence that the ancient conception of meteorology remained unchallenged until almost the end of the century. On the other hand, he acknowledges that many observers were interested in regular recording of weather conditions, not just in anomalous meteors, though he downplays the importance of this. My interpretation aims to give the weather diarists of the time their due, to show how their collective efforts consolidated a sense of the weather as a continuous occurrence that expressed the providential regularity of nature. While I agree that many observers' attention was focused on the atmosphere of their own locality, I shall show how their work fed into the emerging consciousness of the British national climate, allowing for comparisons between the climates of places that were quite distant from one another, such as the homeland and colonial settlements overseas. Building upon Janković's account, I shall connect the activities of the weather observers with the debates about climate and civilization that have been seen as characteristic of the Enlightenment era.

Whereas Janković illuminates the eighteenth century by emphasizing the continuity with what preceded it, other historians of meteorology have looked for precursors of later developments. They have, for example, identified the rise of the discipline with the spread of the "quantifying spirit"—the movement in the last decades of the century toward collecting numerical data and measurements in the sciences.[7] Theodore Feldman has suggested that meteorology assumed a scientific profile only when it got to grips with quantified data and began to discern general trends therein. As the following chapters will stress, however, the keeping of systematic records of the weather, including measurements with meteorological instruments, had already begun a century earlier, in the late seventeenth century, and was practiced fairly continuously from then on. The scientific status of meteorology nonetheless remained contested and insecure, since general laws of the kind that would permit prediction could not be found. As Katharine Anderson has shown in her excellent recent study of the Victorian period, the science of weather forecasting was still viewed as highly problematic in the mid-nineteenth century.[8] It seems to me that the award of scientific credentials to meteorology was considerably more delayed than Feldman and others have claimed, and that the relationship of studies of the weather to their eighteenth-century context should be understood in quite different terms.

The historical development of the science of meteorology is therefore not the central topic of this book. Rather, my aim is to explore how experiences and understandings of the weather reflected cultural change in the eighteenth century. This means that the Enlightenment—viewed as a process of cultural transformation that eventually touched all of Europe and the wider world influenced by it—must feature prominently. The Enlightenment, it has been said, is central to the claim of the eighteenth century upon our attention today, since in many important respects it continues to the present.[9] As the philosopher Mary Midgley has pithily put it, "The Enlightenment is not something safely tucked away in the past, something obsolete that we can patronise. It is where we still live."[10] In subjecting nature to rational investigation, in setting agendas for political and institutional reform, in insisting on the rights of autonomous individuals, and in opening up the public sphere as a domain of communication and free exchange, the movement pointed the way to the modern world. And in experiencing the failures of some of its hopes, in exposing the degree to which human beings remained activated by their passions rather than their reason, in revealing the degree to which cultural traditions continued to hold sway, the thinking of the Enlightenment was also emblematic of the modern age. In the incompleteness and failures of the movement, as well as in its triumphs and accomplishments, we can identify aspects of our modernity. Or, to express the same point in Bruno Latour's provocative formulation, we can recognize the ways in which "we have never been modern."[11]

To suggest we recognize the eighteenth century as a kind of contemporary is not to prejudge what we should find there. On the contrary, it is to endorse an open-minded scholarly inquiry into the rich complexity and diversity of the age. Historians have increasingly been finding that the Enlightenment was not as monolithic as tended to be assumed by those who used to identify it with a unified philosophical program or the outlook of a single "mind."[12] As scholars have uncovered the roots of the movement in localized practices and social conventions, they have brought out more of its heterogeneity. This book relies substantially upon recent scholarship that has shown how natural knowledge emerged as a feature of the eighteenth-century public sphere, in Britain and other countries.[13] The scientific enterprise has been shown to have been fed by local markets for books and periodicals, public lectures, and instruments. We shall see that the public orientation of natural knowledge in this period was essential to its social success but that it also limited its accomplishments (at least from the standpoint of later scientific development). Weather observers remained amateur individuals who largely pursued their own goals; their activities were coordinated only by the mechanisms of voluntary institutions and the market for publications. Scientific theories about weather phenomena coexisted with traditional popular beliefs. Meteorological instruments were often used and interpreted in informal and imprecise ways. The relationship between scientific thought and the "vernacular" thinking of the population at large will emerge as an important theme of this book.

Recent studies of the Enlightenment have shown it to have taken root at different times in different places, and as it spread it conveyed an awareness of the history and geography of cultural differences. Eighteenth-century thinkers themselves recognized the specific cultural character of particular places and populations.[14] In this book, I aim to contribute to this new understanding of the self-conscious diversity of the Enlightenment and thereby to further the current reappraisal of the movement. In relation to the weather, British intellectuals were obliged to acknowledge the continuing hold of traditional beliefs and the crucial role of the passions in mediating the effects of the atmosphere on the human frame. The lessons were brought home to them with renewed emphasis when they confronted people of other cultures living in other climates. Enlightened discussion of the development of civilization attended closely to the influence of the natural environment, acknowledging the ways in which—at the social as well as the individual level—humans were subject to their physical nature. Trying to understand how society developed in particular settings, historians and philosophers discussed how human beings had adapted to a wide range of natural circumstances.

To recover this dimension of the Enlightenment can help to challenge stereotyped views of the movement; it can therefore enhance our appreciation of its legacy in the contemporary world. This seems especially worth doing in connection with ideas about the natural environment. There has been a strong tendency to reject the legacy of the Enlightenment in this domain, partly because the movement has been characterized in a simplistic or one-sided way. It has often been seen as almost entirely pernicious in its environmental attitudes, as an inheritance to be discarded in the interests of achieving a more harmonious balance between human beings and nature. On this account, environmental damage in the modern world is the consequence of a ruthless program sponsored by the Enlightenment to subdue nature by technological and scientific rationality.[15] The agricultural, commercial, and industrial transformations of the eighteenth century are seen as having launched an unprecedented exploitation of natural resources, licensed by thinkers who proposed that nature should submit to human domination. The Enlightenment appears as a campaign to master nature intellectually and technically, to reduce it to administered passivity. In the most general expressions of this view — articulated in the twentieth century by the philosophers Martin Heidegger, Max Horkheimer, and Theodor Adorno — the culprit is the whole program of philosophical rationality initiated by the ancient Greeks. The eighteenth century merely witnessed the climax of a disastrously misguided program of Western reason, which consistently sought absolute sovereignty over nature.[16]

The condemnatory view of the Enlightenment can be traced back to the reactionary movement of Romanticism, and especially to German philosophy of the early nineteenth century. Rejection of the movement's legacy has been central to some of the most ambitious and contentious philosophical programs launched since then. It was also expressed in a more accessible form in the brilliant creative fiction of Mary Shelley's novel *Frankenstein* (1818). With the immediate and profound influence of this novel, the ambitions of enlightened intellectuals in relation to nature assumed the form of a modern myth.[17] Shelley's book was intended as an updated version of the Prometheus story, a retelling for modern times of the legend of the man who trespassed on the domain of the gods and was punished for it. The story of Victor Frankenstein and the creature he makes, the monster that eventually destroys him, was richly ambiguous in its symbolism. But it has become the substance of an extraordinarily persistent and adaptable myth, which has resurfaced regularly over the last two centuries in connection with episodes where scientific reason seems to be set in opposition to nature, and where it is feared that nature will bite back. In the story, Frankenstein's monster is the product of overweening scientific ambition to pervert the course of nature; he personifies that nature as he takes his revenge upon his creator. Shelley consistently identifies the monster with natural forces, including those of the weather. Thunderstorms, lightning strikes, clouds, and mists regularly accompany his appearances. One reason for the success of Shelley's story as a myth is that it mobilizes a religious sense of the breaking of taboos, of the trespass of the profane onto the territory of the sacred, but it translates these notions into a secular idiom. Frankenstein's creative deed seems to have been sacrilegious or at least sinful; but he is punished not by God but by his own creation. Thus, his fate supplies a model for any incident in which it is feared that the product of scientific ambition will turn against its creator as punishment for the act that brought it into existence. People talk of "Frankenstein's monsters" in connection with genetically engineered foods, animal or human cloning, or nuclear weapons.[18] The metaphor is implicit also when people say that natural disasters show nature taking revenge for human trespasses upon its domain. This kind of language ascribes quasi-divine powers to nature, though we are now more likely to personify her as a female goddess than as the masculine monster of Shelley's fantasy. Frankenstein's monster has given birth to the violated goddess, whose vengeful agency is thought to lie behind the climatic disasters of our time.

Though Shelley's tale remains richly suggestive in relation to many situations in the modern world, we should beware of using it too facilely as a key to Enlightenment attitudes to nature. We cannot identify the Enlightenment as a whole with the extreme hubris of Victor Frankenstein, who in some respects represents the Romantic rather than the enlightened ideal of an intellectual. Instead, this book will show eighteenth-century thinkers assuming a more modest stance toward the natural world. In relation to questions of weather and climate, we shall find them acknowledging that humans have never achieved control over nature or even complete knowledge of it. Enlightenment studies of the atmosphere were built upon recognition of the inevitable limits of human reason; they did not manifest rationalistic hubris, still less the incipient totalitarianism that some philosophers have detected among thinkers of the age. Far from expressing the ambition to subject the natural world to human domination, most eighteenth-century writers on the topic viewed human beings as subject to their own inherent nature. Human nature—of which the physical body was the irreducible material substrate—was the inescapable foundation of all society and culture.

As we shall see in the chapters that follow, eighteenth-century ideas about weather and climate expressed a realization that human life remains thoroughly interwoven with the natural environment, notwithstanding the progress of civilization and enlightenment. Civilized society was thought to have been built upon people's natural attributes, even though it was acknowledged as having departed, to some extent, from a natural way of life.

Thus, concerns about how the weather affected people's health and emotions demonstrated the human mind's subjection to the body and its milieu. Scientific instruments invented to measure and predict the weather were interpreted as comparable to—and probably less reliable than—the natural weather signs validated by age-old tradition. Even as civilization brought nature under its sway to a certain degree, it remained vulnerable to atmospheric influences on human health that were often thought to be increasing in the conditions of modern society. In all these instances, eighteenth-century thinkers found nature to be at the heart of their culture, rather than impinging upon it from the outside like some exiled, avenging deity.

This is a lesson worth taking to heart as we consider the legacy of the Enlightenment for contemporary thinking about environmental issues. There are subtleties and ambivalences in eighteenth-century attitudes that are well worth recovering. They counter the claims that the thought of the period set human beings against nature, separated them categorically from it, or set out to conquer and subdue it. Such claims have resulted in an insistence that the Enlightenment legacy should be entirely rejected, that what is needed is a revolutionary break with the past, a wholly new metaphysics, or a new understanding of the human ontological situation. As a historian, I find the notion of such a discontinuous break with the past implausible. The influence of the Enlightenment cannot be wished away or facilely transcended, any more than the whole Western tradition of the scientific approach to nature can be set aside. I also suggest that a more historically nuanced understanding of the Enlightenment can recuperate some of its diverse resources, which are obliterated by interpretations that reduce the ideas of the period to a uniform philosophical program. This can make available to us resources from our intellectual inheritance that can help us think through our own relations with the natural environment.

Michel Foucault—whose engagement with the legacy of the Enlightenment was profound, notwithstanding his reputation as a leading "postmodern" thinker—declared that we should refuse what he called the "blackmail" of summarizing the movement as a totality, to be judged "good" or "bad." The Enlightenment cannot be reduced to a single essence, cannot be evaluated in such a glib way.[19] Nor, of course, can it be rejected in toto, disposed of as if it were an embarrassing article we had inherited from a past we were determined to escape. It may seem perverse to use the British weather as a vehicle for launching such reflections, but I suggest this is as good a way as any to show how we still inhabit the intellectual world that the Enlightenment opened up. For too long, the weather has been regarded as unworthy of serious consideration as a feature of cultural life.

It has been treated as the stuff of jokes and anecdotes rather than of analytical discussion. In this book, I propose to take it very seriously, to use it as a key to the emergence of modernity. This is to connect the apparently trivial with the ponderously serious, to connect something lighthearted with the realm of ideas that the British have usually seen as heavy with the weight of Continental philosophy. My claim is that the way the weather bears upon human life is symptomatic of how nature impinges on human culture, the understanding of which is central to much of scientific and philosophical thought in the eighteenth century and since. We still live in the house of enlightenment, after all, and we can still hear the wind and rain rattling at its windows.

Experiencing the Weather in 1703: Observation and Feelings

They stared at the [clouds] which stuck out like mares' tails, those that look like islands, those that one might take for snow mountains, trying to distinguish nimbus from cirrus, stratus from cumulus, but the shapes changed before they could find the names.

GUSTAVE FLAUBERT · *Bouvard and Pécuchet*

It must however be confessed on this head, that, as our air blows hot and cold,—wet and dry, ten times in a day, we have [the passions] in no regular and settled way.

LAURENCE STERNE · *The Life and Opinions of Tristram Shandy*

WHAT MIGHT YOU SEE if you looked at the sky in the west of England at the beginning of the eighteenth century? One observer, in Worcestershire in early January 1703, saw this:

I remark we had a constant thick & heavy Sea of clouds & close dark nebulous expanse, or Black sad Atmosphere baked in massy clouds, & I could compare ye huge rising body & vast aeriall Load or ye mundane smoak to nothing more than a Diffusion of ye Ocean or steam of some infinite Abyss & what I term in my speciall Language, a Sea-sheet . . . & now we had a Deluge of vapours wich off some exalted eminence, seemed to flow over ye hills & fill ye valleys & invade ye trembling air . . . so that in recompense for my neglect I subjoyne this descant; & note ye year commences wet.[1]

The writer had begun the year by making brief annotations of the daily weather, but at this point—ten days into January—he decided to embark on a much more detailed and descriptive journal.[2] Apologizing for having at first neglected his duty, he resolved henceforth to give an ample narrative of atmospheric events. The diarist did in fact keep up his journal throughout the following year. The result is a small leather-bound volume containing more than two hundred pages of minuscule handwriting. By an accident of fate it survived to our own day, though almost nobody has read it in three hundred years. It includes descriptions of the weather for every day of 1703, a year still remembered for the "Great Storm" of 26–27 November, whose deadly violence laid waste to much of southern England. On the day after the storm, and on many others, the diarist wrote lengthy passages in a richly poetic style. Analogies and metaphors were pressed into service in a strenuous attempt to represent the ever-changing air. What the author called his "speciall Language" was an idiosyncratic response to the challenge of registering atmospheric changes; it gives the document its highly individualistic style. At one point, the diarist suggested a name for his enterprise by comparison with the work of the celebrated astronomer Johannes Hevelius, whose map of the moon had been published in 1647: "I should think my name as immortall by a consummate exquisite raw Atmography, as Hevelius by a sublimer Selenography" (272).

Sadly, however, the author of the 1703 diary has not been immortalized; in fact, we cannot be entirely sure who he was. The most likely identification makes him a young man, recently returned to the family home with a degree from Oxford. Although he left no other known record of his intellectual interests, he seems to have profited from his education, since he mentions a number of prominent natural philosophers of the seventeenth century, including Johannes Kepler, Athanasius Kircher, René Descartes, and Robert Boyle, alongside such classical authors as Aristotle and Seneca. The diary refers to the ideas of these writers on chemistry, cosmology, and theories of the earth. It proposes that atmospheric changes should be viewed as part of a comprehensive cosmological system, in which water and vapors circulate between the air and the caverns beneath the surface of the earth. The "peculiar secret Tours & movements of weather," the author declares, "must be resolved into ye grand architecture of . . . our globe" (357). This perspective enabled the diarist to incorporate elements of popular belief alongside the ideas of learned authors. But it also differentiated his own understanding from the vernacular knowledge of his rural contemporaries. The author was seeking a "philosophical" grasp of the weather that would be distinct from the popular lore with which he was familiar. It is this feature of the diary that makes it particularly useful as a historical document. The 1703 diary shows us its author poised between learned and popular cosmological traditions at the dawn of the Enlightenment.

Another important aspect of the document is the way it joins observations and theoretical ideas together with a record of personal experience. At certain points, the author reflected on his struggles to find appropriate language to represent his very intimate engagement with the atmosphere. Groping for the right descriptive vocabulary, he gave free rein to a poetic impulse to concatenate metaphorical and allusive terms. Words like "sad," "uncomfortable," "lovely," "charming," "smiling," and "cheerful" both name the weather conditions observed and encode the observer's reaction to them. They contrast markedly with the pared-down annotations used by most weather observers from the late seventeenth century on, who saw standardized terminology as the way to compile an "objective" record. The author of the 1703 diary rejoiced in richly evocative descriptive terms that reflected his emotional responses to the states of the air. He also commented directly on the influence of the weather on his health and mood. He described how his bodily humors responded to atmospheric circumstances. He even occasionally recounted feelings of spiritual elation or erotic union with his environment. These seem to have been remarkably intense experiences of identification with his physical surroundings. At times they sound like the emotional outpourings of the Romantic poets; but they also reflect the common early-modern idea that the air has a direct influence on the passions. Such expressions of personal emotions are unexpected in a diary of the weather, which normally demanded the suppression of individual subjectivity. They suggest that, for this author, the weather journal was being used as a means of personal expression, even of spiritual self-development, rather than simply to record what was seen in the sky.

The 1703 weather diary is an unusual document, then, but nonetheless a historically significant one. The author's amalgam of ideas from natural philosophy and popular wisdom was unique. It allowed him to make sense of the atmospheric phenomena he observed and to represent himself as someone intensely aware of the forces of nature influencing him. The quality of the air, as he put it, was the element of his philosophy and descended into his very soul. In this respect, his diary clearly lies off the path that has been recognized as leading to modern scientific meteorology, in which objective records and measurements are systematically compiled by supposedly dispassionate observers.[3] This author felt the weather passionately, and described it as he felt it. Because his journal was so verbose and emotional, he condemned himself to oblivion in the history of meteorology. But for readers three centuries later, his document has a particular value. It suggests that our perspective for understanding past experiences of the weather may have been too narrow. Rather than looking only at those we recognize as having pursued a scientific approach, we should widen our view to encompass observers such as the man from Worcestershire. At the opening of the age of enlightenment, it seems that quite idiosyncratic experiences of the atmosphere were still possible. Realizing this will help us approach the succeeding century in a more authentically historical way. We can come at it by working forward from its origins rather than backward from the more formalized scientific study that followed it. Taking our orientation from the Worcestershire diarist, we shall find that connections between the weather and the bodily passions were widely commented upon in the eighteenth century. As we shall see in later chapters, many other writers discussed how the qualities of the air influenced people's health and mood, though few were as candid as the author of the 1703 weather diary about how they were personally affected.

The "Exquisite Atmography" and Its Author

Given its highly personal content, it is frustrating that the 1703 weather diary was not signed by its author and contains no internal mention of the diarist's name. There are occasional clues as to his domestic and social circumstances. The narrative mentions, for example, features of his residence, such as "our best garden," where on one occasion, in the course of his "usual meteorologic walk," he studied a peculiar frost formation in the mortar of a wall (249). He noted from time to time that observations of the sky had been made from his study window in a "turret" on the house. A few days before the Great Storm, the diarist recorded a tempestuous wind, "so vehement & furious, as to Rock [the] Turret to such a degree, that I could not write well. . . . My candle in [the] study danced & waved as I could hardly see to write" (420). There were other occasions when he mentioned events that impinged directly on the writing of the journal: in December, for example, when the "obstinate" cold "begins to pinch my fingers in writing" (439); or at the beginning of March, when "my inkpot thawed spontaneously about noon" (280). Such "punctual" references are a familiar feature of other diaries from the period. By linking the circumstances in which the journal was written to the events described—referring to the moment of composition within the narrative—they enhance the immediacy and plausibility of the account.[4] But they do not directly point to the identity of the author.

In another publication, I have laid out the trail of evidence that allowed me to make a probable identification of the diarist.[5] An inscription on the flyleaf of the manuscript volume turns out to be misleading, mentioning three men who owned it only after it had left the hands of the original writer. It records the presentation of the book to the Rutland meteorologist and astronomer Thomas Barker by his uncle John Whiston in 1746. Whiston was the son of the renowned mathematician and natural philosopher William Whiston; he worked in the London book trade, where he probably acquired the diary on the death of its previous owner, Henry Bland, dean of Durham Cathedral and provost of Eton College.[6] Barker incorporated observations drawn from the diary in his own weather records, but he did not identify the document's author, who may well have been unknown to him. He did, however, know where the diarist lived, because it is mentioned repeatedly in the document: Edgiock, a hamlet about six kilometers south of Redditch in Worcestershire. There was really only one significant residence in Edgiock at the time — the manor house, occupied by members of the Appletree family from 1656.[7] Unfortunately, the building does not survive, having been demolished around 1890.

It seems likely that the author of the diary had a university education, because he was familiar with the tradition of natural philosophy since ancient times and sometimes used phrases in Latin and Greek.[8] The most likely candidate is Thomas Appletree (1680–1728). He matriculated at Balliol College, Oxford, in 1696 and graduated Bachelor of Arts in October 1699. He was then admitted to the Middle Temple to train for the practice of law, but he soon returned to Worcestershire, where he was managing the affairs of the Edgiock manor house in 1703. It is significant that the diarist mentions visiting Deddington in Oxfordshire in July 1703, since the Appletrees had kinsfolk there.[9] Thomas Appletree was baptized in the parish church at Hanbury, about nine kilometers from Edgiock, on 30 November 1680, and was buried there on 11 August 1728. There is no record of his ever having married, but his will survives in the county record office. It was drawn up a few days before his death, when he was said to be "very sick & weake in Body but of sound & perfect understanding." Appletree asked to be buried in the churchyard at Hanbury, with his cousin George Vernon, the minister there, officiating.[10] This could be the "G. F. Vernon" mentioned in the diary. Perhaps the family connection accounts for the fact that the diarist records traveling to Hanbury to attend church in September 1703, and does not note attendance at the local parish church at Inkberrow.

An inventory following the will records the sale of Appletree's goods as his estate was wound up three days after his funeral. Unfortunately, there is no mention of books or manuscripts. But it is telling that the will mentions as first beneficiaries Appletree's manservant Edward Hollis and his sister Anne Hollis.[11] In the diary, on 28 August 1703, the author appeals to the memory of "old Hollis" concerning the pattern of the seasons. Whether or not this was the same Hollis, the mention of the same name in diary and will establishes a link between the diary and the Appletree family. Taken together with the clues linking the diarist to Deddington in Oxfordshire, to Hanbury church, and to a person called Vernon, this amounts to good circumstantial evidence that Thomas Appletree was the author. It remains possible, of course, that another member of the household or a neighbor wrote the diary. Outside the document itself, Thomas Appletree remains a shadowy figure, with few traces in the historical record. Even within his journal, he rarely showed much awareness of the social world around him. Occasionally, he named individuals with whom he had contact — who, for example, brought him news of weather events or storm damage elsewhere. One of these neighbors, Joseph Bristoll, seems to have shared some of the author's meteorological enthusiasm: the two men watched a lightning storm together in September. On the other hand, there are no references to domestic matters or to public events. Most of the time, the diarist's focus was either on the cosmic or on the physiological level, either on the events he witnessed in the heavens or on their internal effect on his own body, mind, and passions.

The Atmosphere and the Earth

The author of the Worcestershire weather diary was trying to find a literary form adequate to describe the phenomena he witnessed and the intensity of his reaction to them. At one point, he mentioned a previous journal, "my more miscellany confused & immethodicall Diary, before I lighted upon a distinct Tract to enter my annotations in" (403). The existence of a prior volume would explain why the page numbering of the surviving diary begins at 241. Unfortunately, this document has not been found, and in its absence we cannot know what experiments the author had previously made with the language or structure of his journal. In the 1703 volume, he used a system of symbols for recording basic weather conditions: rain or snow, sunshine, the direction and approximate strength of the wind, and so on. Though he mentioned them occasionally, he did not record having used any meteorological instruments. He took no measurements of temperature, pressure, or wind speed. Nor did he show any interest in predicting the weather to come, beyond occasional remarks about certain conditions' "prognosticating" others within a few hours. On some days, a brief annotation using his symbols was all that he recorded. On other occasions, however, he wrote lengthy passages describing the weather and trying to account for it. The ink of the original manuscript varies in tone, sometimes within a single day's remarks, indicating that they were composed in more than one sitting. Sometimes, on the other hand, the remarks for a number of days were written in a uniform ink tone. As the author mentions at one point, his practice was to work from rough notes to compile his commentary on days when the weather seemed to merit extended discussion. This was the procedure by which he composed what he called "my Emphemeris [*sic*] or Historicall Remarques on vicissitudes of weather, with a narrative of its course & Tracing it in its various winding meanders round ye year" (447, 358).

In framing this "historical" narrative, the diarist grappled with the difficulties of using ordinary language to represent the weather conditions he had seen. It was a problem of which he seems to have been keenly aware. "I Remarque," he noted at one point,

> how . . . our Language is exceeding scanty & barren of words to use & express ye various notions I have of Weather &c, I tire myself with Pumping for apt terms & similes to illustrate my Thoughts, & yet must own a deficiency & I cannot invent a Language commensurate to [the] vast & infinite Properties discoverable in meteorology. (357)

Unable to invent his own language, the diarist turned to exploiting the resources of metaphor and analogy. The shifting appearances of the sky ("so changeable a chameleon it is"; 358) and the extraordinary variability of clouds made particularly strenuous demands upon his powers of expression. At one point, he attempted to classify clouds into eight categories, each one based on its resemblance to some commonplace phenomenon, such as the mixture of oils or fats with water. Elsewhere, he suggested metaphorical names for different cloud types: "combs," "palm branches," "foxes' tails," and so on; or he proposed similes, likening them to fleece, cobwebs, crepe, spun wool, raw or finished silk. At times, the metaphors and similes became as bloated and piled up as the clouds themselves:

> Atmosphere loaded & varnished with Bulging, dull swelling Bas-Releive clouds bloated & pendulous. I style them ubera caeli fecunda: sky-cubbies or udders cloudy; they enclosed & stufft ye whole visible Hemisphere in colour like Lead-vapours or a tall Fresco ceiling, or marble veined grotto. (269)

As we shall see later, the diarist was capable of being inspired by clouds to a quite ecstatic degree; he described their appearances in especially inventive terms. In January, rain-bearing clouds were said to be "riding over our heades like vast carracks or hulks of ships in huge Flota of rarified Sea." After they had discharged their rainwater, they "settled in form of some airey-marble" (248). In June, the dark clouds of an impending storm were compared to "swelling up of ye Tide or Torrent of ye overflowing sea." As they gathered ominously at the horizon, they "appeared as a black List, or dismall piece of night stalking on with a solemn pace to envelop us" (326). The richness of the language was justified by the author as a necessary response to the challenge of his task—attempting to convey, as he believed no one had before, the extraordinary variety and vivid impact of meteorological phenomena. He explained that the

> oddnes of such terms must be allowed in a performance of this kind & nature [for the] sake of significancy & [to] exhibit a more naturall & lively idea & image of my meaning, in a subject so little dealt in, & a path so little beaten before me: I embrace with utmost freedom & pleasure any word as conduces to cleer & distinguish my sense or falls in to my purpose. (358)

Clouds—a perennial, if inconstant, feature of the English sky—were usually ignored by later weather diarists, especially those who were compiling quantitative records. They continued to elude categorization until Luke Howard proposed the modern classification scheme for them at the beginning of the nineteenth century.

While rising to the challenge of describing the English weather in words, the Worcestershire diarist also drew upon concepts and terms used in contemporary natural philosophy. Chemistry was prominent among the sciences to which he turned. At times, this was simply in order to find an appropriate metaphor or simile. Thus, an August sky was labeled "antimonial" because it had streaks like the texture of antimony ore (349). The sight of the sun shining through thick clouds was likened to "ye smoak in a laboratory, or Reek of a huge furnace" (290). Although the diarist did not use weather instruments, he referred to laboratory apparatus as models for the global system he believed he was observing in action. The evaporation of water from the sea was described as "Distillation per ventum, . . . ye indefatigable Alchymy of wind & sea wich sublimes & elevates vapours incessantly in ye tall curcurbit of ye Atmosphere" (255). Chemical ideas yielded not just images but also possible explanations for what was observed. For example, the effects of lightning could be explained, it was suggested, "by considering it as a flame maintained by & subsisting in a fluid substance" (376). Wilting in the heat of a July afternoon, the diarist speculated that the "fermented weather" was the result of a "sulphureous secret explosion, . . . a subterraneous emission or exhalation, . . . mixt with some poysonous, Arsenic or illnatured ingredients" (335).

If Thomas Appletree was the author of the diary, he could have learned his chemical terminology during his time at Balliol College between 1696 and 1699. The tradition of chemistry teaching at Oxford had begun in the 1650s, when Robert Boyle had encouraged the German chemist Peter Stahl to give lectures on the subject. Boyle is mentioned numerous times in the 1703 diary, more than any other natural philosopher. His work on cold is referred to more than once, his air pump is mentioned, and his studies of combustion and phosphorus are noted.[12] Oxford chemistry was established on a formal basis in 1683, when Robert Plot was appointed to an ad hominem chair and began to lecture in the basement laboratory of the new Ashmolean Museum.[13] Sir Robert Southwell, a future president of the Royal Society, wrote that he approved of Plot's lectures as a course of study for his son Thomas, a student at Christ Church: "As for the course of chemistry, I like it well enough, though I can hope for him no other benefit by it than an entertainment of the time, whereas to another it proves the directest passage into natural philosophy." [14] Plot resigned from his chair in 1689, but it may well be that instruction in chemistry continued to be available in the late 1690s. The mathematician John Wallis wrote in 1700 that courses of experimental philosophy and chemistry continued "to this time," "from time to time." The Scottish natural philosopher John Keill began to teach mathematics and experimental physics at Balliol in 1694; it is possible that his brother James was teaching medicine and chemistry in Oxford from 1698.[15]

Whether through formal instruction at Oxford or otherwise, the author of the Worcestershire weather diary evidently acquired some knowledge of contemporary chemical theory as well as a smattering of terminology. Perhaps for him chemistry was indeed "the directest passage into natural philosophy." On 15 May, he accounted for the thunder heard that day as an aerial explosion of sulfur and niter vapors released from under the earth: an "excess of solar aestivating heat in these 2 days successive, has raised plenty of sulphurous-nitrous spirits above us, & no [wind] to dissipate or dissolve or blend & mix ye nebulous fermenting Napthaline effluvia very inflammable" (312). This so-called gunpowder theory, which ascribed thunder and lightning to reactions of sulfur and niter in the air, can be traced back to the writings of Paracelsus in the early sixteenth century. The importance of a nitrous component of the atmosphere was emphasized afresh by Oxford research in the 1660s and 1670s.[16] Thomas Willis and John Mayow ascribed a fundamental role to "aerial niter" in sustaining the processes of life, arguing that the explosion of gunpowder was chemically equivalent to the ferment or vital flame that produced motion in animals. In the early 1690s, the same process was invoked by the astronomer John Flamsteed to account for earthquakes, which were thought to result from explosions of nitrous and sulfurous vapors in subterranean cavities. Flamsteed's theory was but one of a number advanced at this time that connected atmospheric phenomena with events in the bowels of the earth.[17] Robert Hooke had been arguing, since the late 1660s, that earthquakes and lightning were caused by explosions of nitrous and sulfurous effluvia.[18] On the other hand, the naturalist Martin Lister proposed in 1684 that they could be accounted for by the ignition of "inflammable breath of pyrites." Sulfurous vapors alone, without a nitrous mixture, seemed to Lister sufficient to cause explosions within the earth or in the atmosphere.[19]

The 1703 diarist echoed some of this speculation, and he aligned himself with its central idea—that meteorological events are fundamentally subterranean in their origin. He wrote of an underground magazine of vapors, the source of all the atmospheric phenomena classed as "meteors" (383). Vapors might be vented through the ground or under the ocean, he suggested. In the latter case, they would carry water up with them into the atmosphere, whence it would eventually fall as rain. The diarist referred to the *Mundus subterraneus* (1665) of Athanasius Kircher as a source for this idea (385, 412), but the German Jesuit was not the only writer speculating along these lines. Ralph Bohun, a fellow of New College, Oxford, wrote in his *Discourse concerning the Origine and Properties of Wind* (1671) about the atmosphere as "one Immense Æolipile," referring to a kind of primitive steam engine in which water was boiled in a sealed vessel and the steam allowed to escape through a narrow nozzle as a powerful jet. For Bohun, jets of water driven up from the sea were the source of the most potent meteorological phenomena.[20] The naturalist and antiquarian Charles Leigh described storms as "occasion'd by Eruptions from the Bowels of the Earth, strugling with that mighty Element [the sea] till they had forced their way through its immense Body." Leigh, in his *Natural History of Lancashire* (1700), also mentioned the aeolipile as an artificial model of the production of winds, as did Plot in his *Natural History of Staffordshire* (1686).[21] The Worcestershire weather diarist thought along similar lines, speculating on the origins of atmospheric vapors in "ye dark infernall dungeons Regions

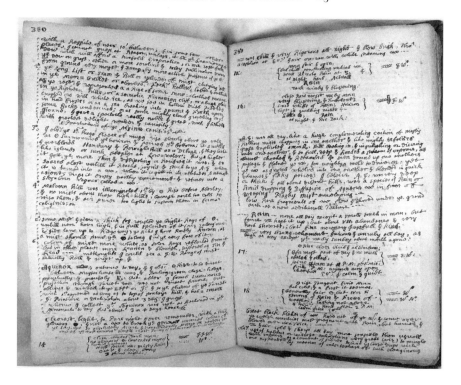

FIGURE 1 · Pages 380–81 of the 1703 Weather Diary. On these pages,
the author begins a series of reflections on the origins of aerial vapors.
Courtesy of West Sussex Record Office, Chichester, West Sussex.

below" (434). He even used the same image as Bohun and Leigh to label
the waterspouts raised by submarine eruptions "a sort of naturall Æolipiles
[which] effect all ye most considerable wonders of meteorology" (254).

The general implication of these ideas was that an understanding of the
processes of the atmosphere would require knowledge of the structure of
the earth from which they arose. Water falling to earth as rain was actually returning to its origin, so that one should imagine "ye internall part
of ye earth . . . [as an] engine to farther ye grand Water-works of nature"
(255). The sky was the great "Atmospherical Theatre" against which the
drama of the "Transmutation & shifting of vapours" was played out (291).
But the backstage area was the ground beneath the observer's feet: "Ye
surface we live upon is but shell, crust, & bore outside of a mighty piece
of mechanism[,] ye case of ye clockwork within" (256). To be understood,
the weather must be decoded as a sign of subterranean events, "a complex
index of all ye movements under ground" (366). This claim was articulated

by the Worcestershire diarist as something of a revelation, a dramatic realization of the degree to which a normally concealed plan of the universe might be uncovered before human eyes. It pointed both to the mysteries of the cosmos and to the way in which they were fundamentally unified, and thus to the possibility of achieving understanding by taking a sufficiently global view.

> We cannot penetrate into ye oeconomy & mysterious inward fabric of this huge machine, we only see ye outside & superficial Plan of it, not ye wonders within. . . . We are ever staring up above over our heades for alterations of weather &c when as ye thing we seek, ye matter we are in quest [of] lies under our feet & our ignorance makes us stumble over it without perceiving so plain & palpable a correction of our stupidity & item to ye Truth; I verily am of this creed, as Meteorology must be establisht & explained by a sound & exquisite Geology or Penetration into ye mundus subterraneus, what we call ye heavens is but ye excrement & refuse of what we name Hell. (365)

Realizing that the heavens were composed of hell's excrement suggested, to the author, the paradoxical nature of philosophical truth. To look below one's feet for the fundamental secrets of the cosmos was the opposite of a commonsense approach. It required the observer to wrench himself away from everyday assumptions about the natural world. In striking this attitude, the diarist drew upon the works of contemporary natural philosophers who had explored how the intricacies of divine design were worked out in the structure of the earth. The 1690s was the decade of the "theories of the earth," attempts to reconcile scriptural narratives of creation and the early history of the world with accounts of the operation of natural causes. The debate was ignited by Thomas Burnet, whose *Telluris Theoria Sacra* began to appear in Latin in 1681, was completed in 1689, and was translated into English two years later. The Worcestershire weather diarist referred at one point to Burnet's well-known claim that the unevenness of the earth's surface was a consequence of the universal flood and a mark of the fallen state of mankind thereafter (389). Burnet's work was challenged by those of John Woodward (1695) and William Whiston (1696), both of whom offered alternative ways to reconcile scripture with the findings of geology, astronomy, and natural philosophy. There was significant interest in these theories in Oxford, where Thomas Appletree could easily have heard them discussed. Edward Lhwyd, Plot's successor as Keeper of the Ashmolean Museum, wrote to John Ray and Martin Lister about the contending theories of the earth, all of which he ridiculed as "wrangling philosophy" that was diverting scholars from solid natural history.[22] John Keill, Appletree's contemporary at Balliol, took issue with Burnet and Whiston in his *Examination of Dr. Burnet's Theory of the Earth* (1698).

The Edgiock diary does not pronounce on the details of any of these theories. It has nothing to say, for example, about issues of scriptural interpretation. However, the author does emphasize the extent to which the cosmos manifests God's design in the architecture of the natural world. "I doubt not God whom wind and seas obey disposes of clouds & Rain to use & benefit of that corner it is directed to," the author notes (260). He expresses confidence that the mechanism of the earth and the atmosphere has been designed

> for wise & suitable ends . . . by ye unerring hand of ye author of nature. . . . Ye infinity of Divine architecture & variety of movements below, ye secret springs & wheeles of this stupendous frame, artifice & admirable composition is to us for ever unseen, unknown, as much above our comprehension, as below our search & indagation. (256)

The diarist appears to have been awed by the magnitude and wisdom of the divine design, and perhaps also inspired by its unity to believe that it could be grasped by the human mind. The weather, if properly understood, could provide a key to understanding the cosmic system as a whole: "So I conclude to be weather-wise one must be All-wise in a manner" (366).

The Worcestershire diarist was not the only writer in provincial England at this time who was musing on the workings of the earth and the atmosphere. As Vladimir Janković has recently shown, meteorology had an accepted place in local natural histories and chorographical writings. Authors who were describing the topography and natural history of their regions would routinely include remarks about local atmospheric phenomena. Leigh and Plot described the weather and meteors in their natural histories of the counties of Lancashire (1700), Oxfordshire (1677), and Staffordshire (1686).[23] Two other writers with similar interests were the Cumberland clergyman Thomas Robinson and the Dorset excise officer William Hobbs. Robinson, whose *New Observations on the Natural History of This World of Matter and This World of Life* appeared in 1696, was rector of Ousby, near Penrith in Cumberland. He observed from the top of Cross Fell, from where he claimed to see both east and west coasts and the circulation of vapors between air and earth. From this vantage point, he pronounced on the merits of Moses as a natural philosopher and the errors of Burnet,

Woodward, and Whiston. Like the Edgiock diarist, Robinson believed that atmospheric phenomena reflected the influence of forces from within the earth: "Vapours and Exhalations are the Perspirations of the terraqueous Globe, and are caus'd as well by the Internal Heat and Fermentation of it, as the External Influence of the Sun." [24] William Hobbs wrote his treatise *The Earth Generated and Anatomized* in 1715, though it was not published until 1981. Hobbs lacked a university education and was almost entirely isolated from the community of natural philosophers. His overture to the Royal Society of London failed to spark interest in his ideas among the scientific elite. He had somewhat the same ambition as the Worcestershire author to grasp the system of the cosmos as a whole, though he based his theories primarily on the earth's strata and the tides rather than on observations of the weather.

These rural philosophers straddled cultural domains that were in some respects growing apart at this time. On the one hand, Thomas Appletree had benefited from a university education that had exposed him to the ideas of some of the leading natural philosophers of the early-modern era. If the 1703 diary was indeed his, then it shows him to have been thoroughly familiar with the latest scientific theories about chemical processes operating in the earth and the atmosphere. These speculations were underwritten by a general view of nature as the realm of a lawgiver-God, a divinity who did not intervene arbitrarily or unpredictably, but whose providential hand obeyed his own regular laws. On the other hand, even educated men living in rural areas were surrounded by people with very different beliefs. The Edgiock diary refers repeatedly to the traditional beliefs of the locals, to sayings, fragments of verse, and pieces of country lore. Uneducated people, according to the diarist, tended to see natural phenomena in a different light. They were more likely to be apprehensive of events such as storms, fearing that they expressed God's anger and his willingness to punish sinners.

The Worcestershire author was generally inclined to view storms as entirely natural phenomena. But he felt keenly the emotional shock of violent meteorological events, and this gave him some sympathy with those who viewed them as omens of divine punishment. This was the pattern of his response to a series of heavy thunderstorms in early September 1703. His first comment on the thunderbolts was that God "keepes such invincible weapons in his hands to avenge his injured honour upon wretched & ingrateful slaves below" (369–70). He invoked a series of classical and biblical references to illustrate this idea, including stories of the Roman emperor Caligula being frightened by thunder, and the Hebrew prophet

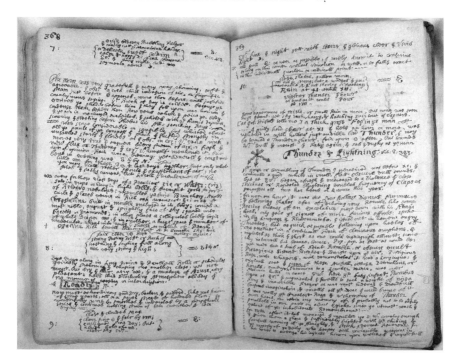

FIGURE 2 · Pages 368–69 of the 1703 Weather Diary.
These pages include the author's annotations on thunderstorms in September 1703.
Courtesy of West Sussex Record Office, Chichester, West Sussex.

Elijah calling down God's thunderbolts to cow the priests of Baal.[25] While
he advanced explanations of these phenomena in terms of natural causes,
he also showed an interest in the fear they engendered among the population at large, a fear he acknowledged sharing to some degree himself. At
one point during a storm, he noted a vivid, luminous formation in the sky
"wich by some might be entred among ye prodigies, & fancy augment into
ye frightfull shape of a Fiery head humane" (376). As the weather cleared,
however, the diarist's reflective, philosophical mood returned. In this state
of mind, he was able to contemplate weather events as part of the divine
plan, confident that fundamentally, they were all part of a good God's wise
design: "Use & application of all these Contemplations ought to terminate
in reverence & awfull attache of deity, converted to morality & devotion. . . .
Take away ye wonder, & ye fear Vanishes; & superstition ceases, destitute
of a supply" (388–89).

His comments on vulgar fears and fantasies reveal the Worcestershire
diarist's position between the learned world and that of popular culture.

On the one hand, he deployed classical and philosophical learning to try to penetrate to the causes of weather phenomena, causes that lay behind (or literally beneath) observed atmospheric changes. On the other hand, like other weather observers since, he found that meteorological study led him into an encounter with vernacular weather lore.[26] He relied on general opinion, as well as on the testimonies of named individuals, to establish that a particular season was hotter, colder, or wetter than any other in living memory. In August, "old Hollis" reported to him the common belief that "every fine, dry summer & plentifull crops are followed with a low, cruel, & Rigorous severe & long winter, a providence, as cattell may spend fodder & convert into soyl for manure" (357). He was willing to side with popular opinion, against the consensus of scholars, in holding that solid meteors did in fact fall from the sky (387). He analyzed the apprehension among the harvest-gatherers on an August afternoon as storm clouds appeared to threaten them: "Ye Dubious & suspicious Æquivocating Ambiguity of weather inclined every body to construe ye worst. . . . Fear, is ye grand Agent, naturall & necessary intention of all human Actions, . . . thô not sole Principle & primum mobile, as Hobbs advances" (352). Here, the diarist stepped from his own educated outlook into the mental world of his rural neighbors, and then stepped back again as he invoked the philosopher Thomas Hobbes's theory of human nature.

The movement back and forth between elite and popular perspectives was one the Worcestershire diarist made repeatedly. Remarking the clouds hovering over nearby Bredon Hill, he noted that this was a sign "providence has given to ye wise & thinking part of men" that rain was imminent (319; cf. 289). One hundred and seventy years later, the omen was recorded as an element of local folklore.[27] A mild, cloudy day in November was said by a relative of the diarist to be an example of "Allhallen-summer" (412). The term "All-hallown summer" could be found in Shakespeare, but it had also long been in popular use for what in the New World would come to be called "Indian summer."[28] Once, the diarist distanced himself from the belief "among [the] Vulgar" that lightning caused beer to go sour (374). But on another occasion, a chill breeze at dawn seemed to bear out the "Countrymans saying" that the break of day was the coldest time (368). Engaging with popular beliefs surrounding the weather, and drawing upon the resources of local lore and memory, the diarist was able to incorporate insights and traditions of the uneducated populace. Like later weather observers, for all the pains he took to distance himself from "vulgar superstitions," he nonetheless relied upon popular beliefs as sources of information. As we shall see in the next chapter, some historians have argued that the cultures

of the learned and the unlearned were undergoing a fundamental separation at this time. Although the man from Worcestershire was capable of
differentiating between his own beliefs and those of his rural neighbors, he
did not perceive an unbridgeable chasm between them.

Clouds in the Head

If the 1703 diary *was* written by Thomas Appletree, then the rusticated Oxford lawyer clearly recognized the importance of vernacular lore and local
tradition when it came to knowledge about the weather. Nonetheless, he was
engaged in an enterprise that had distinct pretensions to learning, a project that he labeled "philosophical." This seems to have referred both to the
form of the record and to its underlying purpose of self-development. The
basic idea of compiling a daily journal of the weather was one that had been
advocated by leading members of the community of natural philosophers,
including Robert Boyle and Robert Hooke. Diaries devoted to routine recording of the weather were kept by members of the Royal Society from the
1660s, following the example of the Accademia del Cimento in Florence.
William Derham, vicar of Upminster in Essex, published his weather journals for every day from 1697 to 1702. The philosopher and physician John
Locke kept up his daily record from December 1691 to May 1703.[29] As we
shall see in later chapters, there were many other contemporaries compiling
meteorological records at the time when this particular diarist was at work.

Nonetheless, the Edgiock observer chose a unique style in which to
write. Other weather diarists suppressed personal details and superfluous
descriptive terms; they reduced their journals to minimalist annotations of
the weather in a standardized format, without any reference to the conditions of the observer. The Worcestershire diarist, on the other hand, called
his text "this Book of Life," "the grand history & picture of my own life"
(358, 447). His work seems much closer than that of other journal writers
to the tradition of spiritual discipline and providential accounting, which
has been identified as an inspiration for diary-writing in the early-modern
period. From the late sixteenth century, spiritual journals had been kept
by devoted Christians (mostly Protestants) to record their triumphs over
temptation and their lapses into sin, and to keep a kind of ledger of divine
gifts and punishments. Tom Webster has argued that this kind of document constituted what Michel Foucault called a "technology of the self,"
"a means by which the godly self was maintained, indeed constructed,
through the action of writing."[30] Foucault argued that the formation of
modern identity was bound up with individuals' self-subjection to forms of

discipline like that involved in keeping an account of one's life. More than most weather diarists, the author of the 1703 text exemplified this model of a journal as an exercise in self-development, a document in which personal identity was articulated through disciplined self-observation.

Choosing to pursue this project through a study of the weather, he also drew upon philosophical traditions that went back to the ancient world. The classical tradition of meteorology had grown up in close connection with philosophical ethics. Knowing about the weather was supposed to be morally improving, especially because a philosophical understanding of the reasons for atmospheric phenomena could dispel the fear they caused among the uneducated. The philosophically prepared individual would be able to subdue the anxiety aroused by thunder, lightning, and other meteors by understanding that they were part of the normal order of nature.[31] The Worcestershire diarist seems to have known this tradition quite well. He sometimes used classical Greek words to describe the appearance of the sky—for example, φασις (*phasis*) for a kind of hazy mist. He referred repeatedly to Aristotle's writings on meteorology, in which earthy and watery exhalations were ascribed fundamental importance as causes of atmospheric phenomena. He quoted a substantial passage from Aristotle's *Meteorologica* to enlist the philosopher's support for the existence of lunar rainbows. And he used the term "medicelestiall" to name the realm of the skies, echoing the Greek idea that what happened there was influenced both by the movements of astronomical bodies and by events in the earth below.[32]

The diarist also reached back to classical tradition when he reflected on people's fears of thunder and lightning. To bolster his own philosophical fortitude in the face of thunderstorms, he turned for encouragement to the Roman poets Horace and Lucretius.[33] The latter had discussed meteorological phenomena in his epic philosophical poem of the first century BCE, *De rerum natura* (*The Nature of Things*). A follower of the Greek philosopher Epicurus, Lucretius proposed that a state of calm suspension of the passions, or "ataraxia," should be the aim of the philosophical devotee. Understanding that violent weather events were not the actions of vengeful gods, but simply natural occurrences, could help the aspirant philosopher along the path to this goal.[34] It was a path the Edgiock diarist seems to have wanted to follow. During the thunderstorms of early September 1703, he referred at length to the Roman Stoic writer Seneca, who had written extensively on meteorology in his work *Naturales quaestiones* (*Natural Questions*), composed shortly before his death in 65 CE. Seneca, the diarist noted, "handles this subject ye most copious large & Rationally, & most Philosophically & happily of any of ye Antients extant" (388). Seneca's reassuring approach to the anxiety raised by violent storms culminated in a pithy reminder that lightning strikes before it can be heard, so that "no one has ever feared any lightning except that which he has escaped." [35]

Whatever philosophical calm the Worcestershire diarist thought he had achieved, however, was severely tested by the Great Storm of 26−27 November 1703—"a most dreadful night for violence of [wind] & rage of it most tremendous & inexpressible" (422). In common with the rest of the population of southern England and the Low Countries, the diarist had his sleep disrupted that night by exceptionally strong winds that up-rooted trees, tore down houses, and sank ships at sea. Natural causes and supernatural forces jostled one another in his narrative as he penned an account of the tempestuous wind. It was "as if [the] whole mass of air were putt into a ferment of convulsions & haunted with Furies, or Exasperated by Aeriall demons & agitated or spurd with all ye Æolian Powers" (422). Trees were uprooted and buildings damaged, he recorded, including the family seat at Edgiock: "All lookt with a decayed, antient Ruinous face, & air of desolation, as some old dropping deserted worn & threadbare man-sion of a Lost & careles owner. . . . With a melancholy aspect did I behold these hostilities of an angry [wind]" (423). The turbulent wind was ac-companied by torrential rain, which led to further damage from floods. A few days later, in early December, he was able to place the storm in a national context, thanks to the availability of newspaper accounts. "Dole-full & Tragicall stories come from all parts of [the] kingdom," he noted, "but I referr ys. narrative to Public papers, wich may be consulted" (427). The same few days' reflection allowed the diarist to suggest a naturalistic explanation for the event, in which he ascribed the storm to a submarine eruption, like a huge cannon, which projected a stream of water and air over the country. The north of England, he explained, had largely escaped its effects because the strength of the jet was diminished at a greater dis-tance from "ye orifice of ye submarine Tube or Oceanic Æolipile" (427).

Perhaps the author of the diary did derive some comfort from giving a philosophical account of the storm that had caused such damage to his house. In this way, he could represent a frightening and violent event as part of the order of nature, having its place in the universal divine plan. He could scarcely have conceived, when he started his journal, that the year would close with such a cataclysm, but the writing of the diary might have helped to make the crisis of the storm more easily bearable. Drawing upon the resources of natural philosophy, he could go beyond the description of such an event to try to explain it. By invoking the forces of nature erupting from beneath the surface of the sea, he reiterated the picture of a cosmic system in which the weather was but an index of more fundamental terrestrial powers. And by narrating his own reflections at a moment of terror and emergency, he wrote himself into the account, reminding himself of the stoical calm in the face of adversity that a philosopher was supposed to attain.

It is notable, however, that the diarist's enthusiasm for his task seemed to diminish after the November storm. The event itself did not attract as much commentary as one might have hoped for, and by the end of the following month, he was confessing that his attention to his duty had slackened:

> I never took less heed, or was more incumbred & totally slurd over my Diary & Ephemeris than today. I could [not] take any notes, or breif abstract for use of my memory & now ye many passages & matters I mett with drove it out of my head; I had allmost lost or mist a day, & left a gap or chasm in my commentories thereby; . . . I lapsed into pristine innocent Gulf of oblivion narrative & this most sweetly prevailes on me & will sooner wean me from this Tiresom & Tyrannic custom of keeping & so laboriously penning this Book of Life, than all ye anxious distasts possible. . . . This I have entered to fairly confess a Defect & failure in nicer observation: dies sine Lineâ. (447)

Philosophical stoicism, it seems, could go only so far. Although the Worcestershire diarist was interested in how a philosophical training could prepare one to remain calm in the face of violent weather events, the November storm appears to have had a genuinely depressing effect on him. His entries for the remainder of the year were fairly perfunctory. And indeed, throughout the diary, he seems to have been more interested in recording his own subjection to the weather than in claiming to be able to withstand its effects. When conditions were good, his spirits were raised and his enthusiasm for keeping his journal was enhanced. On a fine day in August, he noted that the fair prospect "obliges [me] to pay ye gratefull Tribute of a few Remarks upon it, for ye Pleasure I took & received in ye contemplation" (354). More frequently, he characterized his mood as one of saturnine melancholy, to which the damp and misty air played a kind of musical accompaniment. A few days after Christmas, he noted that it was the calm, murky weather of that season that really suited his "Philosophical Quietism" (448). The image of self expressed in the journal was one that was highly susceptible to the weather's influences—not at all a paragon of stoical fortitude.

This note was struck from the beginning of the document, from the first time the author introduced himself as a character in the elemental drama he was recounting. Early in the year, he portrayed himself "taking my usuall meteorologic walk . . . & solacing myself in ye quiet recesses of Philosophicall Reflexions" (249). Thereafter, he related his moods, feelings, and even fantasies to the prevailing weather conditions. A "chill driving Rain" was "a kind of weather as never fails to discompose me" (296). The burning heat of July "made me feint & allmost swoon & even wasted me to the degree of *deliquium animi* [failure of spirit]" (335). A lightning flash made "a naturall impression & terrible surprize on my spirits . . . [and] just exanimated me, so superlative & extraordinary enormous a fright I never was sensible of in all my life" (371). An October day with heavy clouds was a "temper of weather [that] exactly corresponds to my saturnine & quiet melancholy Genius . . . [and] carries a mysterious secret & unknown Conformity to my being" (409). These effects on the observer's body were understood both as instances of sympathy between macrocosm and microcosm and as examples of mechanical causation. An agreeable misty rain was "a soothing Anodyne to my perplexing vexations, & strikes unison to my constitution & falls in patt with my humour"; it was also "weather as chiefly settles the fibres of the brain & ideas even" (343). Sometimes he talked about the particles of air invading the pores of his skin; at other times he invoked astrology, perhaps ironically ascribing his melancholy humor to the influence of the planet Saturn. The diarist was working both with the Renaissance vocabulary of correspondences and harmonies and with ideas of causation derived from the mechanical philosophy as he tried to comprehend how the weather affected him.

At times, the author's sense of communion with the elements reached an ecstatic height that he described as a kind of spiritual rapture. An exceptionally clear day in August "perfectly revived & fed me & I drunk it in at the open windows of the soul"; it "provoked me to soar aloft, & allmost mount above grovelyng mortality, & dart above by force of extasy: the distinct proper & congeniall joys of a Contemplative spirit; the merit & due reward of indefatigable Thought & study" (351).[56] Fine weather enabled the author to overcome the melancholy disposition that was thought typical of a scholar, to attain a vision of ecstatic communion with the atmosphere. On the other hand, a fine misty rain in August "was incomparably delicate & agreeable to my temper[,] a distillation of divine juice; & smoothed my soul all into Raptures, melted me into a sacred & aeternall transport[.] Oh! how happy was I for ye moment, sliding into unknown joys and serenity of mind" (343). A similar day in October had "a sort of passionate Lachrymall serenity, [which] Ravisht me into softest & most amiable contemplations, Tickled & sooth'd me into gratefull content & flusht with silent peacefull solid joy . . . even embraced & folded my soul in the bosom of its own bliss" (402). Clouds, in particular, inspired the diarist to heights of rapture, beguiling, as he put it, "ye whishful chace of my pursuing eies" (346). In January, he described them as "cherishing & cheerfull" (245). In February, he imagined himself scaling the ramparts of castles and cities built in the clouds—the vision he labeled his "exquisite raw Atmography" (272). In August, and again in November, the clouds presented a celestial landscape of mountains and valleys, blissfully recorded by the observer (346, 412). On 3 November he noted, "I spent a good part of the day in admiring & feeding my eager curiosity with feasting my eies & Regaling my cloud-born or Nubigenous Genius, like Ixion engendered of a cloud, I am ever gazing & as it were Returning to my womb" (413).[37]

At such moments, the diarist seemed to be finding inspiration in the clouds for a kind of transcendental experience. Through philosophical meditation, he was drawn upward toward the heavens, away from the trammels of the material body. Inspired by his exquisite vision of clouds, he imagined himself soaring like an eagle: "I ascend on philosophic wings, & build my nest & extend my observatory above ye clouds but heavy dull mortality still checks my soaring & presumptuous flight" (272). He described this ecstatic state as "a Religious & sound sense & secret magnificat to works of God & nature," a "healing influx of spirits, or fluttering expansion of soul," "a partiall Revelation of truth & consequently of God & good & being & happiness" (351). The mood was understood as simultaneously a philosophical vision of the underlying divine unity of the natural world and a response to the influence of cosmic forces that were both spiritual and physical. In marked contrast to the instances when the author claimed a philosophical detachment from the emotional impact of weather events, these ecstatic passages show him to have been highly susceptible to climatic conditions. Indeed, he appears to have relished their effects as a means of achieving a state of spiritual transcendence. It is striking that these experiences were ascribed to the operation of environmental conditions on the observer's body. Unlike certain of his contemporaries, who cited material causes of ecstatic states of mind as a way of dismissing their spiritual significance, the diarist embraced the influences of the atmosphere as a path to transcendent experience.[38]

Furthermore, the use of words like "bosom" and "womb" gave the ecstatic passages a sexual tone. The erotic vocabulary in the accounts of meditative bliss was paralleled by similar language in some of the diarist's descriptions of the natural world. The author's vision of the cosmos was, at times, an explicitly sexual one. The coming of spring, for example, was portrayed in such terms: "Every vegetable strutted in a vivid new-fed green & amiable freshness, swelled & distended with ye vernall succus, like a Plump vigorous face, or semole breasts full of youth & blown up tight & stiff with Longing desires, & flatus of youthful lust" (297). The "blushing new-born flowers" were fertilized by rain that was labeled "spermatic irrigation" or "Balsamic Panspermicall Panacea Juice of Heaven" (306, 270). Spring showers were said to "shed heavens seed . . . on ye earthes longing womb" (283). Paradoxically, the rain was also thought to originate in "the Gulf, bosom & vagina of the ocean; the most secret Recesses, cu—t or Rima magna" (383). Once it was said to have been released by the "copulation & coition of ye two elements," earth and water; on another occasion, it was identified as "milk shed from heaven suckt by each infant plant" (327, 305). The sexual imagery seems to have been confused in its attribution of male and female roles, but it was global in its application: the natural world was represented as an arena of sexual activity on a cosmic scale. And at times when he was particularly aware of the sexuality of the world around him, the diarist reported a sense of erotic communion with the elements. Thus, the vision of spring as an outpouring of sexual activity left him feeling as if "translated or wrapt into Paradise, or Arabia felix," the "cordiall odours invigerating ye wondring soul, & feed[ing] her with unknown delights & rais[ing] every wondring sense into an extasy" (298).

It is worth recalling that if my identification of the diarist is correct, this account was written by a young man of twenty-three, still at this point— and indeed destined to remain—unmarried. It is tempting to interpret incidents in which he found himself transported by the rampant sexuality of the spring, or in which he contemplated a return to his own womb among the clouds, as reflections of sexual frustration or neurosis. Given the lack of information about the diarist's life, however, and the remaining uncertainty surrounding his identity, it is difficult to develop this interpretation in a meaningful way. We simply cannot say what the personal significance of such episodes might have been to young Thomas Appletree, if indeed he was the person who penned them.

It is possible to say, however, that the sexuality of the cosmos was a prevailing feature of both popular and learned beliefs in this period. The diarist's contemporary Thomas Robinson offered sexual explanations of the way metals and vegetation were generated in the bowels of the earth, using terms borrowed from alchemical writers of the sixteenth and seventeenth centuries. In the case of vegetable growth, Robinson suggested, "Seminal Forms or Plastick Souls" were planted in the "warm and moist womb" of virgin matter, from which they drew nourishment to "put forth spungy Strings and Roots." [39] The Worcestershire diarist wrote his own speculative notes on the causes of vegetable growth on the last page of his manuscript, making an explicit parallel with the processes of human sexuality. He sometimes tried to provide a distinguished philosophical lineage for such ideas, as when he invoked Johannes Kepler's concept of a "world soul" to support the idea of marine ejaculations into the atmosphere.[40] But the idea of a sexual basis to vegetative processes was widely diffused in agricultural society at large, and it seems likely that popular tradition also extended the notion to geological and meteorological phenomena. As well as their personal significance for the author, then, the diarist's passages of erotic identification with the natural world seem to be indicative of his acquaintance with both popular and learned systems of belief. His discourse moved repeatedly between the two registers—employing everyday language, for example, to describe the wind as "preservative, & Pickle of ye air," and then shifting to a series of learned terms to designate it "ye meteorologic Architect & Vis Plastica, Figurator, Regulator, vitall active soul, spiritus agens & operator of ye Atmosphere" (312–13).

While the diarist's passionate embrace of the sexuality of the cosmos reflected both learned and popular beliefs, it is not easy to reconcile it with his ambition to achieve philosophical detachment. His ancient precursors, Lucretius and Seneca, had sought to calm the passions rather than to rouse them; they saw philosophical knowledge of the weather as a path to tranquility, not to ecstasy. The writer of the 1703 diary was also contravening the ethical norms prevailing among natural philosophers in his own time. The leading intellectuals of the era insisted that control of the passions was a precondition for the exercise of reason or for making objective observations of the natural world. Various techniques of regimen, bodily exercise, and mental discipline were employed by such seventeenth-century natural philosophers as Boyle, Newton, and Locke, to subdue the passions and thereby fit the senses for observation and the mind for reasoning.[41] Writers such as Henry More and Meric Casaubon denounced the "enthusiasts" who mistook internal agitation for external sensations or genuine spiritual inspiration. In 1662, More had pointed an accusing finger at those enthusiasts who were susceptible to "the mere change of weather and various tempers of the Aire." Four decades later, John Trenchard condemned fantasists who were "actuated wholly by their several Complexions, Constitutions and Distempers, which often make them *Ixion* like, embrace their own Clouds and Foggs for Deities."[42] Had they known of him, they might have had the Worcestershire diarist in mind. They would surely have seen him as having surrendered to his own internal clouds and fogs, succumbing to what Robert Burton had identified as "windy melancholy."[45] Anyone who found spiritual inspiration among the clouds would have seemed to them to have allowed his mind to become obscured by the mists of his passions.

It is hardly surprising, then, that the Edgiock diarist's passionate and expressive style did not subsequently become the norm for meteorological journal-keeping. His "speciall Language" never became the public idiom of weather observation. What he called "my lonely genius" (402) remained isolated indeed. His record remained unpublished, and even half a century later, Thomas Barker could make little of it. As we shall see, subsequent inquirers into the connections between weather and health generally excluded any emotional or autobiographical content from their diaries. If they mentioned their own health at all, it was only when an indisposition led to a lapse in the record. They paid no attention to their own bodily conditions, in some cases keeping up their journals to within a few days of their deaths. They used their bodies instrumentally — in other words, as conditions of observation rather than as subjects of it. Following the recommendations of such individuals as More and Boyle, they disciplined themselves to be passionless and objective. They kept their subjective experiences out of their narratives as a way of assuring readers that what they reported was unbiased by personal involvement.

The Worcestershire author, on the other hand, had chosen to develop his journal as a means of personal expression and self-exploration. He enthusiastically monitored the effects of weather conditions on his own state of physical and emotional well-being. He happily exhibited his own sensitivity to the disturbing effects of meteorological conditions, which other diarists silently concealed. He recorded feelings of rapturous union with cosmic forces, when he contemplated the coming of spring or lost himself in elation gazing at the clouds. The author's flights of fancy among the clouds appear both as responses to the influence of fine weather upon his "spirits" and as moments of mystical and erotic bliss. Presumably, his desire to record these feelings provided one motivation for keeping the diary, as important perhaps as the opportunity it offered for speculative excursions into natural philosophy. The diarist recognized the potential of the genre for compiling a record of how the physical environment impinged upon his body and his mind. This was to make a weather diary into a unique kind of "technology of the self," in which a literary form originally used for purposes of spiritual self-development was adapted to monitor the subjective effects of the physical environment. The diarist catalogued his own subjection to the forces of the atmosphere, while also tracing his attempts to master them by philosophical reflection. He showed himself finding spiritual resources for self-development in the very material circumstances to which he was subjected.

This gives his peculiar document a special historical value. Part of its fascination is that it represents a path not taken in subsequent science. As Michel Serres has noted, clouds have remained largely external to mainstream scientific thought, because "one must not have one's head in them, nor have them in one's head."[44] Contemplation of clouds still risks being equated with mental turbidity. The Worcestershire diarist represents an alternative to the style of objective observation that was already emerging among natural philosophers in this period. While sharing with his better-remembered contemporaries an interest in the natural causes of atmospheric events, he cultivated a mode of experiencing the physical environment quite at odds with theirs. He saw himself as resonating with aerial influences that conveyed genuine philosophical insight. The quality of the atmosphere, he wrote, was "symphonicall to my Genius, . . . adapted to Grave deeper Thought: element of my Philosophy . . . & descends into my soul" (395). For him, the meteorological realm was not simply a domain for objective study, but a medium of contemplative transcendence and spiritual rapture. Reading the words of this anonymous—and hitherto forgotten—individual, we are reminded of the possibility of experiences of the natural world quite other than those historically sanctioned as scientific. His text speaks across the centuries of an especially intimate relationship between an individual human being and the material environment within which life is lived.

The Edgiock diarist was not the only person to try to cultivate such a relationship with the environment. Some of his weather-related preoccupations resurfaced among British writers a hundred years later, during the period of Romanticism in art and literature. Then, Luke Howard turned his eyes toward the clouds, developing a classification system for them that has stood the test of time. Howard captured the attention of his contemporaries more effectively than had the Worcestershire diarist. The German poet Goethe was inspired to write letters to him. John Constable incorporated his clouds in his landscape paintings.[45] A few years earlier, Richard Townley had compiled a journal of his year's residence on the Isle of Man, displaying many of the same concerns as the Edgiock diarist. Townley recorded weather conditions daily, worried about how they were affecting his health, and observed how the customs and sayings of local people reflected them. As one historian has written, "The weather is actually the main character in [Townley's] book and the one that determines

FIGURE 3 · Engraving of cloud types according to Luke
Howard's classification. The plate, by Lewis, accompanied
Howard's paper "On the Modifications of Clouds," in
Philosophical Magazine 16 (1803): 97–107. Courtesy of
Science and Society Picture Library, London.

events."[46] His meticulous recording of the character of the atmosphere on
each day and his obsessive rendering of meteorological conditions in psy-
chological terms recall those of the Worcestershire writer almost a century
earlier. Townley strove for a similar attunement of his bodily state with the
physical environment around him, and understood it similarly in quasi-
spiritual terms.

In a sense, then, the Edgiock diarist was a Romantic before his time. But the themes he developed were not entirely absent from the intervening decades of the eighteenth century. With the advantage of acquaintance with his record, we can make out features we might otherwise miss in the age of enlightenment. His confidence in the providential wisdom of the divine design, for example, underwrote the efforts of many other weather recorders who did not discuss it explicitly. And when anomalous atmospheric events occurred, such as epidemics or the summer haze of 1783, observers still pointed to the role of exhalations from the interior of the earth. At such moments, they were also reminded of the attitudes of uneducated people, and they tried—as the Worcestershire diarist had—to balance philosophical detachment with acknowledgment of the emotional impact of such events. The idea that the weather influenced health and emotions remained part of the mentality of the age. It was a perennial topic of popular and learned opinion, and one of the domains where the two intersected. It cropped up when people discussed the behavior of weather instruments, like the barometer and thermometer, which seemed to mirror bodily reactions to the states of the air.[47] It was the subject of serious deliberations by medical writers; and it provided material for vernacular sayings, for conversational exchanges, and for witty remarks by novelists. Our encounter with the diarist from Edgiock has prepared us to acknowledge these themes in the century that followed his lifetime, even though the period remained entirely oblivious of the startling observations of this isolated and forgotten figure.

Public Weather and the Culture of Enlightenment

If we come into a more contracted Assembly of Men and Women, the Talk generally runs upon the Weather, Fashions, News, and the like Publick Topicks.   JOSEPH ADDISON · in *The Spectator*

THE WORCESTERSHIRE WEATHER DIARY of 1703 was an entirely private document. The author seems to have compiled it solely for his own purposes; there are no indications that it was ever intended to be published. But that year also saw significant public discussion of the weather in Britain. Many other writers shared the diarist's preoccupation with observing atmospheric phenomena and trying to explain them in philosophical and theological terms. A flurry of publications followed the Great Storm in late November, reflecting the terror of the event and its appalling cost in lives and property. The storm was widely interpreted by the authors of pamphlets and sermons as an act of divine punishment for the sins of humanity. Traditionally, extraordinary events of this kind had been thought of as direct interventions by God, interruptions of the normal order of nature. They had often also been seen as omens of war or political change. By the beginning of the eighteenth century, this view was being challenged by those who urged a more "philosophical" approach to the weather. They argued that all atmospheric events, even extremely violent ones, were entirely natural. They claimed that God's providence took the form of upholding the regular laws of nature, laws that might ultimately become known by systematic study. The storm brought this debate into the open arena of public discussion; it forced people to think about the role of extraordinary weather phenomena in the British climate and how they related to God's providential care of the nation.

In this period, ideas about the British weather were being reshaped by a variety of cultural and social forces. The 1703 storm struck a particularly hard blow against shipping at a time when the British sense of national identity was beginning to emerge, identified as it was with naval power and overseas trade. In addition, cultural changes that we associate with the beginnings of the Enlightenment shaped the meanings ascribed to such events. Some intellectuals argued that anomalous weather should not monopolize the attention of the public. To highlight occasional violent irregularities, they suggested, was to disparage the goodness of God, which was expressed in the overall regularity of the national climate. Most of the time, the weather varied only within moderate limits; it behaved with a uniformity for which people should be grateful. During the eighteenth century, this notion of the national climate came to the fore. The British weather came to be seen as an example of God's providential goodness to the island's people, his benevolence in bestowing upon them conditions that fostered the growth of agriculture and commerce. The national climate was represented as bound up with the character of the people and a condition of the progress of their civilization. As such, it was considered a suitable topic for conversation in polite circles—an uncontroversial aspect of common knowledge, untainted by partisanship or superstition. Joseph Addison, in the daily periodical *The Spectator* (1711–12), wrote that the discourse in polite companies of men and women frequently turned on the weather as a public topic.[1] By becoming a public matter, comparable to fashion and news, the weather had entered into enlightened discourse. It had been tamed or "civilized," rendered an appropriate attribute of a nation that prided itself on its reason, refinement, and sensibility.

The polite or civilized view of the national climate reflected the British people's most favorable view of themselves, as a prosperous, industrious, and enlightened nation. But this is not the whole story of the British preoccupation with weather in the period. In practice, people's talk about the subject was not always particularly refined. Certain kinds of talk—that of illiterate or rural people, for example, and also gossip and proverbial speech among the educated—reverted to older patterns. Even polite conversation would frequently resort to oral lore and vernacular traditions that predated the spread of enlightened culture. While "vulgar superstitions" were regularly denounced in civil discourse, traditional oral wisdom was also widely respected, as we saw in the case of the Worcestershire diarist. When it came to weather forecasting, in fact, proverbial lore gave the best advice available. Furthermore, the weather itself continued to throw up violent and seemingly inexplicable phenomena, which were widely reported and commented upon. Their status as public events was considerably enhanced by the growth of printed news media in the eighteenth century. In this respect, the 1703 storm established a pattern of widespread reporting and diverse interpretation that resurfaced in connection with later events such as the summer haze of 1783. These anomalies were a reminder that the weather remained stubbornly unpredictable and sometimes dangerous, notwithstanding the efforts of enlightened investigators to subdue it by scientific reason. Such experiences stirred passions among the populace at large that the educated elite were inclined to regard as superstitions. Events of this kind therefore created anxieties about the prospects for cultural progress. In its occasional extremities and anomalies, the weather showed the British people an image of their society that was less than completely enlightened.

The Great Storm in Public Debate

Probably nobody in southern England slept undisturbed through the night of 26–27 November 1703. The weather had already been blustery for several days; a gale the previous week had nearly blown the Edgiock diarist off his feet as he returned from a neighbor's house. But the wind really started to blow hard in the afternoon of Friday, 26 November. As people cowered in their houses, they heard it wreak destruction around them throughout the night, uprooting trees, tearing tiles off roofs, and bringing chimneys and church steeples crashing down. The wind was so violent that many thought the earth itself was quaking. Churches were damaged throughout the southern counties; if their spires survived, they often had the lead stripped from their roofs. Among the fatalities was the bishop of Bath and Wells, killed with his wife in bed by a chimney that fell through the ceiling. The Eddystone lighthouse in Plymouth was swept away, with its architect—who had placed too much faith in its stability—inside. Hundreds of windmills were destroyed, their machinery broken by the wind or set on fire by friction. Millers tried feeding grain to the mills to slow them down, with limited success. In London, the tempest reached its peak of intensity in the early hours of Saturday morning. Later that day it abated, and a dazed populace began to take stock of the appalling damage left in its wake. Around 120 people were found to have died on land, with many more injured. Damage to buildings in the capital was said to rival that caused by the Great Fire of 1666.

The fatalities were far worse at sea, raising the overall death toll to an estimated eight thousand. Fifteen naval ships and an uncounted number of merchant vessels were lost, many torn from their moorings. What were normally places of refuge became death traps, as ships were dashed against harbor walls and onto the southern shore by the wind. A tidal surge in the Bristol Channel swamped others. Much of the naval fleet had returned to home waters for the winter, encountering stormy weather already along the way. Many vessels sought shelter in the Thames estuary, where they turned out to be in danger of being driven onto sandbanks. Ships in the Downs, off the east Kent coast, were blown eastward to founder on the Goodwin Sands. On the morning after the storm struck, wrecks and stranded men were visible from shore on the sands, which are exposed at low tide. Most drowned as the tide rose inexorably, though the town of Deal mounted a rescue operation that brought a couple of hundred ashore. Up to one thousand naval men died in this one incident, and more probably perished in merchant vessels. In the late twentieth century, archaeologists uncovered remains of the drowned sailors on the Goodwin Sands, poignant reminders of the suffering experienced on that dreadful night.[2]

The Edgiock diarist was not the only person struck with melancholy as he took in the extent of the destruction. In Sussex, eighty-three-year-old John Evelyn, the foremost advocate of arboriculture in the nation, felt particularly for the trees brought down by the storm. He was saddened beyond words by the damage to his own estate, which was "not to be paralleled with any thing happening in our age."[3] The storm entered the national memory as an occasion of unrivaled devastation. An anonymous author who catalogued five hundred years of storms, hurricanes, and earthquakes insisted that "the like has not been known in the Memory of the Eldest Person living in *England*, or any European Nation."[4] Thirty years later, it was remembered as "the most terrible Desolation of the kind that ever was known in the Memory of Man."[5] It was still referred to as the most powerful storm recorded in the British Isles at the time of the "hurricane" of 16 October 1987. The event has earned a permanent place in the historical record, still regularly invoked in news reports of storms three hundred years later.

In fact, the Great Storm's place in history was secured partly by the work of pioneer journalists who made it a news event on a national scale. Special-issue broadsheets, such as *The Amazing Tempest* (1703), gave details of the damage to property and stories of the people who had been killed. A compilation entitled *The Storm* (1704), edited by Daniel Defoe, collected accounts of the deaths and destruction from all parts of the country.[6] Defoe issued a call for contributions in the *London Gazette* on 2 December 1703, "to preserve the Remembrance of the late Dreadful Tempest." He received hundreds of letters and printed a rich selection of the most graphic narratives. Like the other pamphleteers, Defoe emphasized what the Edgiock diarist called "dolefull and Tragicall stories," accounts of individuals who had died or narrowly survived. The underlying assumption of the published accounts was that the storm was a "remarkable and signal ... Judgment of GOD on this Nation."[7] The general moral message was driven home in a variety of ways in the stories of particular individuals: the man crushed by a collapsing chimney as he spoke crossly to his wife, the maid dug alive out of the rubble who thanked God for her deliverance, the couple who saw their baby killed but were themselves spared, and so on. It was only by holding to the idea that divine action was implicated in the fate of these individuals that some sense could be made of the disaster that had struck them.

No writer on the storm argued that God was *not* ultimately responsible for the event. Everyone agreed that it was a "prodigy," an extreme and anomalous occurrence. What was in question was whether it was a "natural" one or not. Some authors—but not all—felt it was necessary to defend the storm's unnatural quality in order to uphold God's role in it. This argument relied upon the common supposition that the winds were tools of God's direct action. One anonymous pamphleteer, who believed that the deity had been particularly angered by the depravities of the London theatre, specifically denied the validity of naturalistic explanations, such as "that the storm was nothing but an Eruption of *Epicurus's* Atoms; a Spring-Tide of Matter and Motion." According to this writer, the tempestuous wind had acted abnormally, and "when Natural Agents act in a strange, unusual manner ... this is from the Lord."[8] In other words, the storm could be recognized as an immediate agent of God's will because it was a departure from the normal course of nature.

This was also Defoe's position. He saw the storm as "the dreadfulest and most universal Judgment that ever Almighty Power thought fit to bring upon this Part of the World." He discounted the notion that the event could be fully explained by natural causes, insisting that "where we find Nature defective in her Discovery, where we see Effects but cannot reach their Causes; ... Nature plainly refers us beyond her Self, to the Mighty Hand of

Infinite Power." Philosophers, ancient and modern, were ultimately clueless when it came to tracing the causes of the wind, which had "blown out the Candle of Reason, and left them all in the Dark."[9] God withheld such knowledge from human minds, because he retained control of the winds to execute his judgments in the world. At least in connection with this dreadful tempest, Defoe was willing to go against the verdict of the philosophers, from Lucretius and Seneca to modern times, who had tried to reassure human beings confronting nature's terrors. On such occasions, he suggested, it was entirely appropriate to be fearful.

It took some time for an alternative, more philosophical view to assert itself. Members of the community of natural philosophers who commented on the storm's effects struck a rather different tone from the compilers of shocking narratives, but were reluctant to engage directly with the theological underpinnings of the journalists' accounts. From Essex, William Derham reported to the *Philosophical Transactions* of the Royal Society, "I shall not weary you with a long History of the Devastations, &c. but rather some particulars of a more Philosophical consideration." He recounted his own observations from Upminster and those of Richard Townley, a correspondent living near Burnley in Lancashire.[10] Both had stoically continued recording barometer readings and rainfall measurements throughout the episode. Derham also placed the November storm in the context of the weather earlier in the year, noting that the damp and mild season had built up vapors in the atmosphere and raised "nitro-sulphureous or other heterogeneous matter, which when mix'd together might make a sort of Explosion." For further development of a naturalistic explanation, he deferred to the astronomer Edmond Halley, an established expert on the causes of winds, who had "undertaken the Province of the late Tempest."[11] A similar report was sent to the Royal Society by the Dutch microscopist and cloth merchant Anton van Leeuwenhoek. He had also watched his barometer closely in Delft, noting that the mercury had never been so low in the tube as when the wind was strongest. When the mercury began to rise, however, he remarked to those with him "that the Storm would not last long; and so it happened." Also trying to maintain a degree of philosophical detachment from the popular mood was the Sussex farmer John Fuller, who sent in a note recording that salt spray had been blown ten miles inland by the wind. He reported that the trees were coated with a salty white rime, and that he could "scarce perswade the Country People, but that the Sea water was blown thus far."[12]

Derham signaled how a naturalistic explanation of the storm might go, but it was difficult to articulate this view publicly in the face of the extreme violence of the event and the shocking magnitude of the suffering.[13] People preferred to see their fate as resting in God's hands, regarding his actions as sometimes inscrutable but always ultimately just. As the author of a poem published shortly after the tempest put it:

> For Thou, O Lord, 'tis only Thou dost know
> Whence the Winds come, whither, and why, they blow.[14]

The prevailing view was that the storm was a heavenly "wonder," one of those events traditionally seen as "preternatural" deviations from the normal course of nature. Many such happenings had been witnessed in the preceding decades, especially coinciding with the military and political events of those turbulent times. A sixteenth-century work by William Fulke, republished in 1670, proclaimed that "the first and efficient cause" of all atmospheric phenomena "is God the worker of all wonders." He classed violent thunderstorms, large hailstones, whirlwinds, auroras, and strange cloud formations, along with comets and shooting stars, as "meteors," since ancient authorities like Aristotle had located them all in the middle heavens between the earth and the moon. Fulke noted that comets "betoken (say some) Wars, Seditions, Changes of Commonwealths, and the Death of Princes and Noble men." [15] Comets that appeared in the 1660s and 1680s were invested with political meanings by many, and the weather accompanying them was scrutinized. When a violent thunderstorm interrupted the second day of Charles II's coronation ceremonies, in April 1661, commentators scrambled to counteract what they anticipated would be a general tendency to interpret it as a negative omen. In 1688, when Charles's brother and successor James II was dethroned by the arrival of the Dutch prince William of Orange, there was widespread celebration of the "Protestant wind" that had ushered William and his army across the Channel.[16] In 1692, an earthquake in London was seen as another "prodigy" or wonder, and was probed for its moral meaning. An anonymous "Reverend Divine" wrote that such marvels "exceed the ordinary course of things, and are above the usual Laws and Power of Nature." Marvelous phenomena were often set in the sky to communicate the divine omen to all who saw them:

> The sudden and unaccountable changes which are sometimes observ'd in the Air and other Elements, the strange and amazing Tempests, Storms, and Thunders, . . . Alterations in the Heavens, strange Appearances of the Sun and Moon . . . [are] set in the fair and spacious Theatre of Heaven as the fittest place to represent those Divine shews to the view of all.[17]

The 1703 storm was not assigned a specific political significance; being very general in its effects, it was more plausibly represented as a judgment on the moral corruption of society at large. But the custom of giving meteorological wonders a political interpretation continued into the second decade of the eighteenth century, though it was met with rising skepticism among the intellectual elite. The transition to the Hanoverian dynasty and the Whig parliamentary hegemony that supported it was marked by such events. Queen Anne's death in 1714 was followed the next year by a total solar eclipse, which, although predicted, was viewed with apprehension by some. Halley organized observers throughout England to send in reports of what they had seen; he published a map of the path of the moon's shadow across the country and a prediction of where the next one would fall in 1724.[18] In 1716, an extraordinary aurora was seen after the execution of the rebel Jacobite lords Kenmure and Derwentwater. "Lord Derwentwater's lights" were interpreted either as tributes to the rebels' patriotic resistance to the Hanoverian succession, or as the fires of hell welcoming them to their damnation, depending on the observer's political persuasion.[19] Halley, on the other hand, proposed that the lights in the night sky were caused by celestial matter circulating in the earth's magnetic field.[20] He and William Whiston campaigned against the notion that these heavenly wonders signified divine displeasure at the new Whig regime. They earned denunciation by an anonymous pamphleteer as "the Lucretiuses of the Age, . . . Modern Epicureans and Libertines" for offering naturalistic accounts of the phenomena.[21]

Even while public fascination with these events continued, however, a philosophical program to "naturalize" them had been gathering strength since the 1660s. Many intellectuals believed that all purported prodigies should be reduced to natural occurrences. Learned opinion assumed that miracles had ceased, and demonic interference in the natural world was no longer suspected. It seemed theologically preferable to emphasize God's general providence, manifested in the uniformity of nature's processes, rather than his particular interventions. The trend coincided with the gradual elimination of the category of the preternatural as its contents were redistributed between the expanding realm of natural phenomena and the ebbing domain of the miraculous.[22] Samuel Pepys — diarist, naval administrator, and fellow of the Royal Society — thought it was "a foolery" to worry about a thunderstorm during the king's coronation. Astronomers like Halley who calculated the motions of comets expected that predictions of their appearance would lessen the apprehension surrounding them. Halley also took the lead in mapping the patterns of winds across the oceans, attempting to tame even those notoriously capricious powers of the air.[23] As we shall see in a later chapter, contemplation of the barometer led virtuosi and philosophers to debate the possible causes of winds, including solar heating and the effects of vapors that might increase the density of the air. Speculation about atmospheric "effluvia" originating beneath the surface of the earth also contributed to a tendency to see natural causes at work. In the 1690s, theorists of the earth, including Thomas Burnet and William Whiston, argued that one could even give a naturalistic explanation of something like the biblical flood, while recognizing that it nonetheless served God's purposes.[24] The point was that God had acted through the normal forces of nature, even while bringing about an event that was unique in the earth's history.

This perspective was shared by some commentators on the 1703 storm, who, though happy to draw out the moral implications of its fearful divine judgment, were also inclined to acknowledge that it had natural causes. The author of a historical survey of atmospheric wonders that was published the following year insisted that they manifested God's power, "amazingly represented in Inferior Things to make the proudest tremble, when he reproves Man for Sin." But he also promised to give the natural causes of a long inventory of meteors, including "Winds, Storms, Earthquakes, Blazing-Stars, many Suns and Moons seen at a time, dreadful Apparitions in the Air, fiery Dragons and Drakes, circles around the Sun and Moon, Rainbows seen in the Day and Night, . . . Thunder, Lightning, Vapours, Mists, Dews, Hail, Rain, Snow, and Frost, and Lights that lead People out of their way in the Night."[25] All these terrifying and amazing phenomena were due to the normal forces of nature. Winds, according to this writer, were just elemental air with admixtures of earth and water. A few years later, a more resolutely naturalistic account was given of a thunderstorm that struck Richmond in Surrey on Whit Sunday, 1711. The author, who witnessed the frightening event firsthand, turned for reassurance to the natural philosophers who had accounted for lightning by sulfur and niter exploding in the air. The gunpowder explanation showed how wrong were those who regarded thunderstorms "as extraordinary and immediate Judgments from Heaven for the Wickedness of those who suffer 'em: For there is nothing in all this which supposes or implies any immediate Interposition of God."[26] Of course, in a general way, God was responsible for everything that happened, but he had established the powers of nature at the creation of the world and thereafter allowed them to run their course. It was incredible to believe that he would interfere with the natural order to administer punishment to a single individual.

A more systematic statement of the naturalistic approach was given by John Pointer in 1723. Pointer was chaplain of Merton College, Oxford, and rector of Slapton in Northamptonshire; he wrote a number of works on history and antiquities. In his *Rational Account of the Weather* (1723), he called for the weather to be reduced to natural law, as Newton had tamed the motions of the planets and comets by making them accountable to the law of gravity. The key was to grasp that "Natural Causes do Naturally (i.e. according to the settled Order and Nature of things) produce Natural Effects." [27] To reinforce the point, he drew upon works by Derham and Halley on atmospheric vapors and the causes of winds. In the second edition, he included a reference to Bernard Annely, whose theory proposing to explain the winds was presented to the Royal Society in 1729. In the final pages of his book, Pointer took on the author of one of the more sensational accounts of the 1716 aurora, a pamphleteer who had asserted that the phenomenon could not be accounted for by the ordinary course of nature and so must be regarded as a "prodigy." To Pointer, such a claim was a cover for ignorance and little short of sacrilegious: "The Extraordinary Power of GOD is to be accounted very Sacred, and not to be touch'd or expos'd for Our Pleasure or Conveniency." [28] God was dishonored by those who claimed he was always working miracles, just as much as by those who denied that he had ever done so. The appropriate theological stance was not to fear the capricious actions of the deity but to respect his ordinary providence as shown in the design of the natural world.

There was a political element to Pointer's argument as well as a theological one. Those who represented strange meteors as omens risked arousing fear among the general population. What could be gained by making people "(as it were) Planet-struck with a Panick Fear?" The consequences of this "kind of Enthusiasm" could only be politically pernicious, stirring up the populace to agitation or subversion. The same results would follow from reports of armies fighting in the air, a phenomenon long recorded as an omen of battles or wars. As far as Pointer was concerned, what was seen was simply clouds and lightning. The claim was fanciful and foolish, albeit "seriously mentioned by some of our Old Historians, and too credulously believ'd by the Vulgar." Rather than a genuine appearance in the sky, this was a creature of the brain, a product of people's own "Superstitious Imaginations." [29] Pointer spoke for the preponderance of elite opinion at this time, which was becoming increasingly scornful of the popular fascination with wonders and marvels. By the early eighteenth century, the interpretation of these as omens had come to be seen as a type of enthusiasm, a disease of the imagination with potentially dangerous effects in society at large. As the "polite" social elite distanced itself from the culture of the general population, it fostered an image of the "vulgar" imagination as particularly susceptible to such delusions. Genteel people feared that indulgence of these excesses could disrupt the prosperity and social stability they were increasingly enjoying, threatening a return to the brutal political and military conflicts of the seventeenth century.

Denigration of the popular passion for wonders and marvels went along with a diminishing emphasis on "particular providences," those occasions when God directly intervened in the lives of individuals. Although early-seventeenth-century Puritans had tended to regard God as immediately involved in their individual lives, this outlook had receded in the second half of the century. Good or bad weather was no longer regarded as a personal sign of divine favor or disfavor. What was emphasized instead was God's "general providence," the uniform benevolence discernible in the design of a universe governed by regular laws. On this level, the divine will could be clearly perceived and merited human admiration. A calm appreciation of the order of nature was considered the appropriate way for the enlightened to pay tribute to their creator.[30]

As they were relegated to the domain of popular superstition, prodigies and marvels were stripped of the significance they had previously held in seventeenth-century natural philosophy. Since the medieval period, natural philosophers had been fascinated by singular wonders of nature, whether shining stones, unicorns' horns, or animals with strange birth defects. Francis Bacon had pointed out the significance of such "sports" as instances of "nature erring," when natural forces had been diverted from their normal course into the peculiar forms of preternatural phenomena. Wonders of this kind were collected with gusto by Bacon's followers, including many of the members of the early Royal Society. Singular facts were in some respects the prototypical facts of nature, the prized specimens of what seems in retrospect a rather uncritical kind of empiricism. But as the category of the preternatural was squeezed out of existence, monsters and marvels ended up as simple anomalies, divorced from any general significance. They were still captivating, reported both in the general press and in learned journals, but their fundamental meaning was unclear. They remained facts of nature, but stripped of the theological and philosophical significance they had previously enjoyed.[31]

These developments affected the treatment of extreme and unusual weather phenomena. Such phenomena did not vanish, nor could an event like the tempest of 1703 be ignored. They remained worthy of the attention of journalists and even of natural philosophers. In fact, the tradition of studying prodigious meteors enjoyed a considerable afterlife in the eighteenth century; the phenomena continued to be reported both in the works of local naturalists and in such August publications as the *Philosophical Transactions*.[32] But such anomalies resisted the attempts of enlightened intellectuals to explain them. As late as the 1730s, a Dissenting preacher was reminding his congregation that the 1703 storm had still not been explained by even the most ingenious natural philosophers.[33] At the same time, suspicion of the merely marvelous and a new interest in the uniformity of nature had led investigators toward a different focus of study. Their search was directed toward finding underlying patterns in the weather that could reduce it to regularity, if not to uniformity. Even something as extraordinary as the Great Storm could eventually be assimilated to the long-term trends of the British climate, viewed as a manifestation of God's general providence. The reporter of the 1711 Richmond thunderstorm denied that God was immediately active on that occasion; but he reminded his readers at the end of the account that it was appropriate to praise the deity for the temperateness of the English climate, which prevented this kind of event from occurring very often.[34] When a freak tornado struck the Sussex coast in May 1729 and penetrated several miles inland, the author of a published account declared that he wanted to lead readers to a "philosophical" understanding of the event by situating it in the context of the normal weather in the area.[35] Ultimately, even the 1703 storm was remembered not for the damage it caused, or even the casualties, but for the fact that it was such a singular and extreme departure from the normal equanimity of the national climate.

Providence and the British Climate

The storm of 1703 was a "public" event in more than one sense. It was reported in pamphlets and newssheets pouring from the London presses, at the very epicenter of the storm's impact. The volume of ephemeral and periodical publications was rising rapidly at this time. Newspapers were still something of a novelty, but they were beginning to seize the attention of readers in the capital and beyond. Defoe was one of those who saw the opportunity presented by the catastrophe to feed the growing public appetite for printed news. He compiled his collection of storm stories, assuring readers that they could trust its accuracy notwithstanding the outlandish events described. He used the same narrative conventions that had been used by seventeenth-century natural philosophers: naming witnesses and detailing the circumstances in which remarkable phenomena had occurred. As he explained, a narrative included in the collection, "tho' it may be related for the sake of its Strangeness or Novelty, . . . shall nevertheless come in the Company of all its Uncertainties, and the Reader [be] left to judge of its Truth."[36] Authenticated in this way, "strange but true" occurrences, including extraordinary weather events, found their place in journalistic practice.[37]

The storm was also public insofar as it was a disaster on a specifically national level. Although its effects were felt as far away as Scandinavia, it was regarded by the British as having struck particularly at their national security and prosperity, especially through the damage inflicted on the naval and merchant fleets. As the preacher of one sermon declared, "It's a *National Stroke* that we are now Lamenting."[38] A month after the event, Queen Anne issued a proclamation, mobilizing the church hierarchy and local officials to collect charity to support the widows and orphans of dead mariners. The tempest was also taken by some as a call to moral reform of the nation, a stern reminder that standards of public morality had declined since the God-fearing days of Puritan dominance. Because of the storm's public status, however, it was impossible to uphold the view that God had directed it to punish particular individuals. Its effects were so evidently widespread and apparently indiscriminate that it had to be admitted that the innocent had suffered along with the guilty. Those who regarded it as God's will acknowledged that such a general chastisement would inevitably touch the blameless as well as the sinful. This was a strong argument for identifying the event with God's general—rather than his particular— providence. It suggested that the divine mission had been delegated to subordinate forces that did not discriminate among the victims, even if (in the view of some commentators) those forces operated outside the normal order of nature.

The 1703 storm remained an anomalous event, albeit a highly public one. Most of the time, when the weather was less violent and extreme, it was a more straightforward matter to represent it as part of a uniform order of nature. Its public status in fact encouraged people to interpret it in this way. The connection has been noted by historians as typical of the scientific knowledge constituted during the Enlightenment. The institutional structures of what has come to be known as the "public sphere" often fostered a sense of the natural world as homogeneous and regular.[39] Printed publications, the proceedings of formal and informal societies, and lectures and conversation in the new urban gathering places provided space for the creation of a form of natural knowledge that was supposedly available everywhere and to everyone. The condition of this knowledge was taken to be a nature that was the same in every place and for every observer. Just as enlightened society was represented as open to all, so the natural world was expected to be available to be known by anyone. Just as Europeans anticipated that enlightened progress would spread globally, so nature was expected to be everywhere the same. The social order, supposedly grounded on universal features of human nature, found its counterpart in a natural order that would appear universally homogeneous. Of course, things did not always work out as smoothly as this intellectual "uniformitarianism" suggested, especially not where the weather was concerned. Anomalies and exceptions confronted attempts to force the natural world into patterns of uniformity; general laws were often difficult to apply in particular situations. But the vision of reducing natural occurrences to a uniform order of laws was a powerful ideal that inspired much of Enlightenment science.

Study of the weather, which obviously affected everyone, was one of the areas in which investigators were guided by this ambition. John Pointer introduced his own account of how weather phenomena could be reduced to rational law with a reminder of the universality of the human experience:

> Air (or the different Temperature of it, by which we mean WEATHER) is one of the grand Concerns of Mankind. 'Tis what affects all Sorts of People, Young as well as Old, Sick as well as Strong. Insomuch that even those very persons that for want of Health are confin'd in close Rooms, feel either the Good or Ill Effects of the *Weather*. The *Air* being like Food, the better, the more refreshing. Hence it is that the Sick Man is inquisitive what *Weather* it will be; and the Healthful, when he is to take a Journey, is willing to consult his *Weather-Glass*. And even those of the *Fair Sex*, are unwilling to stir abroad unless the *Weather* be like themselves, and they like the *Weather*.[40]

Everyone had an interest in understanding the weather and predicting its behavior if possible. It was a public experience even for those who were shut up at home. Individuals composed themselves to go out into it, as they put on a public face to emerge into society. Men and women found that its humor or temperament conformed with their own, or not, on particular days. It was common to all individuals and specifically social in that it was experienced and interpreted collectively.

Knowledge of the weather was also public in that it was cultivated by institutions and circulated through the medium of print. Many of the diarists who labored to compile systematic records of the weather were motivated by a sense of obligation to the public. They often submitted their records to learned societies or had them printed; sometimes they made use of preprinted forms, which were already circulating among observers in the late seventeenth century.[41] Connections of this kind gave them a sense of participation in a larger community of investigators, something that may not have been important to the Edgiock diarist but was for many others. Thus, systematic recording of the weather really started in Britain when it was promoted by some of the leading lights of the Royal Society in the 1660s. Robert Hooke published his "Method for Making a History of the Weather" in Thomas Sprat's *History of the Royal Society* in 1667, laying out the format a daily journal should take. Robert Boyle persuaded John Locke to start a weather diary in June 1666. Portions of it were published in Boyle's *General History of the Air* (1692), which Locke edited for publication after the death of its author. When Locke sent a subsequent installment of his journal to Sir Hans Sloane in December 1700, he wrote, "This I know that I did not keep this register for my own sake alone." [42] William Derham's meticulous journal was offered to the world at large in the Royal Society's *Philosophical Transactions,* as was part of Richard Townley's from Lancashire and Robert Plot's from Oxford.

In the 1720s, a renewed effort was launched to coordinate weather recording by James Jurin, secretary of the Royal Society. He circulated a printed invitation to prospective observers, giving advice about appropriate instruments and procedures. Dozens of weather recorders were inspired with a sense of public-spiritedness to contribute to this project. Journals were received from many places in Britain, continental Europe, and North America. Most of them languished unpublished in the society's archive, as Jurin found himself submerged in data and unable to create a comprehensive synthesis.[43] In the following decades, however, weather records continued to appear in the *Philosophical Transactions* from Surrey, Cornwall, Devon, Somerset, Rutland, Northamptonshire, Dublin, Maryland, Georgia, South Carolina, Hudson's Bay, and Madeira. By 1774, the society was beginning to compile and publish its own weather record, using instruments kept at its London premises. The next decade saw the publication of journals from Bristol, Edinburgh, Somerset, Montreal, the Coast of Coromandel in India, and the Coast of Labrador in North America. In the 1780s, British efforts in this direction were rivaled by new institutional initiatives in France and Germany that established the enterprise of weather recording on a more systematically organized basis.[44] From the vantage point of the end of the century, previous efforts to keep records looked sporadic and disorganized. But the efforts of dozens of weather diarists prior to the 1780s should not be disparaged. By sending their work to a body such as the Royal Society, they confirmed their ambition to contribute to the public fund of knowledge. They were encouraged by a sense of participating in a collective enterprise that was building up a picture of the climate of Britain and its overseas territories.

The *Philosophical Transactions* were far from the only outlet for such publications in Britain. *The Gentleman's Magazine* (a general periodical rather than a learned journal) began in 1751 to include monthly accounts of the weather sent in by the London physician John Fothergill. For four years, Fothergill gave discursive summaries of the variations in temperature and pressure, the quantity of rainfall, and the diseases that prevailed in the capital.[45] Shortly thereafter, an annual diary of the weather began to appear in the magazine, contributed by (among others) Thomas Barker in Rutland and Gilbert White in Hampshire. Many physicians who were interested in the effects of climate on health incorporated weather journals in their own medical writings. They included Charles Bisset and Clifton Wintringham from Yorkshire, Lionel Chalmers from South Carolina, George Cleghorn from Minorca, and William Hillary, who began practice in Yorkshire before relocating to Barbados. Other medical practitioners provided diaries of the weather in their published travel accounts, including Hans Sloane (later president of the Royal Society), who went to Jamaica in the late 1680s, and Tobias Smollett (better known as a novelist), who visited Nice in the 1760s. Books were even published that consisted solely of weather records, such as Benjamin Hutchinson's *Calendar of the Weather for the Year 1781* (1782), Hayman Rooke's *Meteorological Register* (1795), and William Bent's *Eight Meteorological Journals* (1801). Considering that they often made for extremely dull reading, the volume of such publications in the eighteenth century is remarkable. They testify to the sense among observers that they had something valuable to offer to the world, that systematic weather records could potentially be of significant public benefit.

As a record of the weather began to be compiled in the public domain, a complementary discourse about the British national climate began to circulate. The notion of "climate" was an ancient one, originally meaning simply a zone of latitude on the globe and later extended to cover aspects of the physical environment of a place, including its atmosphere. Some clichés about the British climate had been commonplace in the learned tradition since the writings of Tacitus and Julius Caesar. Classical writers had already recorded that the island was wretchedly damp in comparison with Mediterranean lands, though lacking in extreme cold. But in the course of the eighteenth century, the British came to see their national climate in a much more favorable light, appropriating for themselves the temperate ideal that the ancients had assigned to the Mediterranean. Regular recording drew out the routine features of the British atmosphere, rather than fo-

FIGURE 4 · James Gillray, "Dreadful Hot Weather" (1808). One
of Gillray's series showing typical British characters in various
weather conditions. Hot weather clearly does not suit the John Bull
type. Courtesy of The Lewis Walpole Library, Yale University.

cusing exclusively on extreme peculiarities like the 1703 storm. It tended to
"normalize" the weather, making it a quotidian process that went on all the
time and not just when dramatic events drew special attention.[46] Viewed in
this way, the national climate appeared generally benevolent—both moderate overall and gently variable in temperature and precipitation. Both its
temperance and its changeability came to be seen as assets to the agriculture, commerce, health, and character of the nation.

This new view of the climate was bound up with the sense of British
national identity that was emerging in the period.[47] After the union of
England and Scotland in 1707 and the subsequent Hanoverian succession,

Great Britain was constituted as a political unit, and a corresponding ideology of national identity began to appear. The nation's weather was tied to its physical setting, as an island in the midst of an ocean, and was seen as integral to its destiny as a maritime and commercial power. Clement conditions, with plenty of fertilizing rain, sustained the country's agricultural productivity. The prevailing winds powered oceanic navigation in an age of sail, contributing to the expansion of commerce and empire. The weather was also thought to have shaped the character of the people, their temperament having been hardened by a climate that was often bracing but rarely extremely harsh. It was believed that the national character benefited from the stimulus of frequent atmospheric change, which made people more active and independent-minded than those who lived in placid or tropical regions. Joseph Addison mused repeatedly on the subject in his essays in *The Spectator* (1711–12), touching on the influence of the British weather on fashions and conversation, health and mood, gardening and trade. His colleague Richard Steele wrote in his own periodical, *The Guardian* (1713), about the unforeseeable variations of the British weather and their value for toughening the moral fiber of the inhabitants. Addison juxtaposed the temperateness of the island's climate with the command of overseas trade that brought tropical fruit to its door:

> Nor is it the least Part of this our Happiness, that whilst we enjoy the remotest Products of the North and South, we are free from those Extremities of Weather which gave them Birth; That our Eyes are refreshed with the green Fields of *Britain,* at the same time that our Palates are feasted with fruits that rise between the Tropicks.[48]

Thus, the Edgiock diarist was not the only observer of the skies who concluded that God disposed of wind and rain to benefit the places they were directed to. Wind-swept and rain-lashed it might be, but Britain had harvested the winds to send its ships overseas and relished the fertilizing effects of its rainfall. Better to be a beneficiary of this temperate climate and the commercial prosperity it nurtured than to suffer under tropical heat or arctic cold. In the decade after Addison and Steele, Jonathan Swift's *Gulliver's Travels* (1726) satirized this aspect of national pride. When pint-sized Gulliver preposterously boasted to the giant king of Brobdingnag about the greatness of his nation, he began his discourse by dwelling at length "upon the Fertility of our Soil, and the Temperature of our Climate." [49] Gulliver's real-life compatriots were equally convinced that they were climatically blessed.

FIGURE 5 · James Gillray, "Windy Weather" (1808). Another
of Gillray's series on weather and character. This figure shows
fortitude in the face of windy conditions on Hampstead Heath.
Courtesy of The Lewis Walpole Library, Yale University.

The historian and topographical writer John Campbell summed up the
consensus view in his *Political Survey of Britain* (1774):

The climate, though we sometimes hear it censured, as being subject
to frequent and considerable Alterations, is, upon the whole, both
temperate and wholsome, insomuch that we seldom stand in any
Need of Furs to defend us from the Severity of the Cold in Winter, and have more seldom Reason to complain of any insupportable
Heat in Summer. If therefore our Weather be, as is commonly al-
leged, in general less steady and serene than in some other Countries

FIGURE 6 · James Gillray, "Delicious Weather" (1808). Another
from Gillray's series. Here the national character finds its ideal
weather conditions, associated with agricultural fertility,
good health, and flourishing wildlife. Courtesy of The Lewis
Walpole Library, Yale University.

of Europe, it is not so sultry in one Season, or so rigorous in another.
We are subject in a smaller Degree to Storms of Thunder and Lightning; to long piercing Frosts, and deep Snows; and though we have
a full Proportion of Rain, in Ireland particularly, yet it falls moderately, and not with such Weight and Violence as to produce sudden
and dangerous Inundations.[50]

For Campbell and other writers, the beneficial consequences of the British
climate were to be found in the character of its people — in their energetic

initiative, their willingness to explore other parts of the world, their resistance to political tyranny, and their ample share of creative genius. Countries with more serene climates, according to Campbell, tended to support indolent populations of "Bigots or Drones." Of course, the British people were liable to their own airborne diseases, but what country was free of them? Taking a comprehensive view of the climatic situation, Campbell summed up, "We cannot but acknowledge the singular Bounty of Providence in that respect."[51]

That Irish writers such as Swift and Steele contributed to this glorification of the British climate is indicative of the part played by Irish Protestants and the Anglo-Irish elite in the formation of British national identity. Scottish authors also were usually happy to identify themselves as British after the Act of Union, and especially after the defeat of the Jacobite insurrection in 1745; a number of them commented on the medical effects of the national climate. At the same time, Irish and Scottish intellectuals took pride in the meteorological conditions of their own countries and expressed this through compiling weather journals.[52] Irish records went back to William Molyneux, who kept them on behalf of the Dublin Philosophical Society from 1684 to 1686. James Simon's observations in the same city in the early 1750s were published in the *Philosophical Transactions*. Later in the century, the chemist and meteorologist Richard Kirwan reviewed the history of weather observations in Ireland and published his own journal of conditions in Dublin in the 1790s.[53] Scottish records began with Andrew Hay, living near Biggar in Lanarkshire, who made daily annotations of the weather in his diary for 1659 and 1660.[54] A century later, the Edinburgh Philosophical Society published a journal of the years 1764–76. The most dedicated Scottish observer of the last part of the eighteenth century was a woman, Margaret Mackenzie of Delvine, Perthshire. Few women are known to have undertaken the task of weather observation in this period, but Mackenzie did so with great dedication, compiling a meticulous record of daily temperatures at her home from 1780 to 1802.[55]

The Quaker physician John Rutty, who recorded the weather in Dublin from 1725 to 1766, drew conclusions specific to Ireland and made comparisons with the situation on the British mainland. He also concurred with the providential view of climate that formed part of the new sense of British national identity. Rutty was born in Wiltshire in 1698 but went to Ireland to study medicine and settled there to practice. At the end of his career as a weather observer, he penned a justification of the enterprise in the second volume of his *Essay towards a Natural History of the County of Dublin* (1772). He acknowledged that rapid changes of conditions in Ireland made it difficult to draw out general patterns from the record. Nonetheless, prolonged observation could in fact identify some seasonal regularities, notwithstanding the cavils of "Lazy men" who complained that the weather was entirely capricious: "Those who have diligently attended to these operations in nature, will scarcely give up as Chimerical all attempts to discover something regular and periodical therein." To do so would be to turn one's back on divine providence, which had in fact bestowed a benevolent climate on Ireland, as disciplined observation had shown. Rigorous study disclosed "the footsteps of divine Wisdom and Goodness, presiding over these seemingly irregular operations." It could therefore help inculcate the proper attitude of respect for the deity, correcting the "too frequent, not to say wicked exclamations we hear against the inclemency of the Climate, our changeable, and particularly our moist and windy weather, . . . which are owing to a want of attention to this branch of natural history." [56] Rutty seems to have thought that compiling systematic journals, or "histories," of the weather would stop people from complaining about it.

On the British mainland also, the providential benefits of the national climate, including its health-giving properties, were said to emerge from systematic study. The physicians William Hillary and Clifton Wintringham agreed that rainy weather was often healthy; the British people appeared to flourish in damp conditions.[57] Nor was it the case that wind was always bad: Pointer remarked that it depended on the direction from which it came. A westerly breeze was usually healthy; the wind was harmful only when it turned atypically to blow from the north or east.[58] Other writers agreed that winds coming off the Atlantic Ocean were harmless or even beneficial. The situation of Britain in the midst of the sea was generally believed to contribute to the healthiness of its air. Toward the end of the eighteenth century, fashionable people took to visiting seaside resorts to bathe in the waters and imbibe the refreshing breezes. Promoting the Sussex coastal resort of Brighthelmston (later renamed Brighton) in 1782, Dr. Loftus Wood touted the health benefits of its onshore winds. The doctor assured his readers that the Brighton population was so hearty that "the picture of health seems painted in the face of each individual." [59] In those days, visitors to the seaside did not expect to sunbathe; they flocked to the coast to breathe the bracing air.

Medical writers did, of course, admit that the British population often became ill; indeed, they usually took disease rather than health as their subject. Those who monitored the ebb and flow of diseases among the population noted that changes of season frequently ushered in incidents of sickness. Unseasonable weather—summer cold or winter warmth—was also fingered as a risk factor.[60] John Fothergill noted that "it seldom happens that there is any remarkable increase of mortality, without some very sensible change in the temperature of the air preceding it."[61] This made the population particularly susceptible to the frequent changes of the British weather. One commentator suggested that the Englishman was "the weather-cock of the creation," since "you may find him in different humours in several parts of a variegated day."[62] The notion that the English were particularly susceptible to melancholy because of the gloominess of their climate circulated widely in continental Europe.[63] Nonetheless, Fothergill insisted that foreigners were wrong to think that the air in England had "something in it extremely pernicious." Rather, the English people had "abundant cause to be satisfied" with their climate, since it was more temperate than that of any other country.[64] Even if seasonal variations of weather did tend to be risky to health, they would generally be confined within the moderate limits of the temperate norm. And the more rapid fluctuations that occurred on a daily basis were often thought a positive benefit of the British atmosphere. The physician John Arbuthnot, author of the influential *Essay concerning the Effects of Air on Human Bodies* (1733), thought that rapid changes of air temperature and pressure provided people with a stimulating "sort of Exercise."[65] William Falconer, an Edinburgh-trained physician who wrote about the influence of climate on national character, observed that the diurnal mutability of the British weather stimulated the mental alertness of the population, while its general temperateness subdued the passions and thereby fostered good judgment.[66]

Falconer even suggested that the beneficial influence of the British weather could be detected in the way it stimulated the inhabitants to engage in intellectual pursuits like his own. Nor was he the only writer to ascribe British scientific accomplishments to the influence of the climate.[67] The Sheffield physician Thomas Short, who kept weather records for almost four decades and wrote extensively about climate and popular health, saw this kind of research as an index of the degree of enlightenment his society had attained. In non-European regions, Short complained, "the barbarous Natives" had no inclination to keep such records, and even in countries where learning existed, "the Generality of People have been too idle to collect such Histories."[68] Systematic inquiry into the national climate was represented as a token of industriousness and refinement, a sign that the country had pulled itself up from what was thought of as "barbarism" and was taking responsibility for the influence of the physical environment on the welfare of its population.

In this way, the public enterprise of recording the weather came to be identified with cultural progress. Those who were studying the British weather saw their activity as a manifestation of refinement, exhibiting an aspect of nature that was itself consistent with enlightened values. A climate that was providentially regular and moderate—one might say "civilized"—seemed appropriate to a nation that prided itself on its accomplishment of civilization and enlightenment. Refinement or politeness was more than a matter of superficial decorum for the enlightened elite; it was fundamentally important as a cultural marker that set them off from the mass of the population. Polite knowledge of the weather was therefore set in opposition to beliefs identified as "barbarous" or "vulgar," for example the visions of armies fighting in the air that John Pointer had castigated as primitive superstitions. In the same vein, John Rutty bemoaned the fact that "in the last Century it was . . . a prevailing Opinion among the Vulgar, that the Winds were in some measure, under the direction of infernal spirits." Such fallacies were to be displaced by recognition of "the superintendency of a Providence in these seemingly irregular commotions of our Atmosphere." Rutty also denounced the idea that the moon had an influence on winds or rain; superstitions of this kind, he claimed, "have no better a Foundation than heathenish Idolatry . . . [and] cannot stand the test of the growing light of Christianity and sound Philosophy." [69] His investigations were designed to vanquish popular ignorance with the weapon of true knowledge of the atmosphere and the underlying regularities of its behavior.

Other weather diarists reflected similarly on the role of their research in combating the errors of popular belief. They declared that one of their aims was to rescue information about the climate from enthusiasm and exaggeration, unreliable memory and vulgar gossip. At the beginning of the eighteenth century, Samuel Say, vicar of Lowestoft in Suffolk, recorded that he had begun to compile a weather diary, "to be able to contradict some common & groundless Observations and Superstitions." The agricultural writer William Marshall wrote in 1779 that he hoped to see the subject rescued "from the hands of *vulgar Error*." [70] Gilbert White, vicar of Selborne, wrote in 1776 of his own inquiries into weather and natural history that they were a counterweight to "superstitious prejudices . . . too gross for this enlightened age," such as those held by the "lower people" who lacked the advantages of a liberal education. Only those whose minds were challenged to take a broader view could escape the grip of primitive beliefs, "sucked in as it were with our mother's milk." [71] White gave examples of surviving superstitions he had witnessed: magical rituals to heal infirmities or defeat witchcraft that were still being practiced among his rural neighbors. Against these he set his observations of wildlife and the cycles of the seasons, all of which manifested the providential wisdom of God's design. The weather of each season was indispensable to the flourishing of crops and wild plants and to the welfare of the animals and birds that fed upon them. Accordingly, he maintained a record of the weather for thirty-five years, noting the readings of his barometer and thermometer daily, and remarking on how animals and plants coped with the daily conditions. For White, the weather journal was an appropriate part of the natural history of Selborne. He discerned in his record another dimension of God's magnificent handiwork, and a convincing argument against "those who complain about the weather, [showing] that it is generally seasonable for the productions of the earth, and that they complain without cause." [72]

When the weather seemed to depart from its providential regularity, however, White worried that people might forget God's goodness to them. Then, old superstitions could return. The peculiar haze of the summer of 1783 raised these anxieties acutely. It was, he wrote, "an amazing and portentous" season, "full of horrible phenomena." The haze lay everywhere, undisturbed for several weeks from late June to early August. The sun was darkened to the color of blood. At times, the heat was so intense that meat rotted before it could be eaten, even on the day it was killed. White recorded that "the country people began to look with a superstitious awe at the red, louring aspect of the sun; and indeed there was reason for the most enlightened person to be apprehensive." In nearby Fyfield, his brother Henry noted that "ye superstitious in town and country have abounded with ye most direful presages and prognostication." [73] Throughout the country, newspapers reported the murky skies and ensuing violent thunderstorms. They also recorded the apprehensions of the people and episodes of popular panic.[74] Literate observers seem to have had one eye turned to the sky and the other warily monitoring the reactions of their contemporaries. Turning for succor to literature, Gilbert White found a passage in Milton's *Paradise Lost* that seemed applicable, alluding as it did to "a superstitious kind of dread, with which the minds of men are always impressed by such strange and unusual phænomena." [75] Distressing events such as this called into question his faith in the providential regularity of the natural order and in the degree to which his society had become truly enlightened.

Twentieth-century meteorologists have been able to trace the summer haze of 1783 back to a dust and gas plume discharged by the eruption of a volcanic fissure in Iceland. They have even mapped the pattern of atmospheric movements that ushered the volcanic cloud over the British Isles and the European continent.[76] Eighteenth-century investigators assumed that the atmospheric anomaly had its origins under the earth's surface, but at first they looked in the wrong direction—toward earthquakes in Sicily and Calabria thought to have vented gases from subterranean reservoirs.[77] Two years later, Benjamin Franklin suggested that the Icelandic volcano was a possible cause, but his proposal does not seem to have been generally accepted.[78] All inquirers agreed that the event was an unprecedented one, unparalleled either in the memory of living witnesses or in the records compiled by weather diarists. It was a thoroughly public occurrence, reported in dozens of newspapers and the subject of numerous reports to learned societies, but it clearly upset public expectations of the national climate built up over the previous decades. One newspaper labeled the event nothing less than a "universal Perturbation in Nature."[79]

As White's reaction showed, the social implications of such an episode were distinctly disturbing. It reminded the British people that their climate was not entirely temperate and regular, the gift of a uniformly benign providence. It also reminded them that unenlightened attitudes might easily surface among the population at large. While it was no longer respectable to regard weather wonders as miraculous or demonic interventions, the educated elite worried that that was exactly how the masses would interpret them. Fearful portents were supposed to have been banished from the landscape of eighteenth-century learning, but they kept returning to agitate the populace. These uncivilized features of the weather—intrusive and unwanted guests in the public sphere—threatened to overshadow the sunny landscape of enlightenment.

Conversation and Weather Lore

The summer haze of 1783 seemed to highlight differences within British society. The literate elite, self-consciously professing politeness and refinement, scrutinized popular reactions to the event with notable anxiety. They felt they had freed themselves from the fear of portents and wonders, but worried that such superstitions would reappear among the populace at large. Their own unease about the occurrence itself was compounded by disquiet about its socially destabilizing effects. They worried about what might follow if the masses were agitated or panicked. At a moment like this, the enlightened middle classes realized that their beliefs and values were not shared by the mass of the population; the cultural gulf between them and the remainder of society loomed wide.

In fact, many historians have argued that this gap had been getting wider in the course of the eighteenth century. As bourgeois affluence increased, it found outlets in new cultural forms. Members of the social elite took to reading novels and periodicals, patronized the visual arts, enjoyed new sports and entertainments, and improved the architecture and landscape surrounding them. In pursuing these avocations, the literate and affluent distinguished themselves from the way of life of the bulk of the population. They aspired to the world of fashion and sensibility, enjoying the experiences of the metropolis and the new urban resorts; they withdrew from the rural festivals and rough sports that were still popular among the masses. At the same time, new ideas and values—those identified with "politeness," "refinement," or "improvement"—were articulated in opposition to the beliefs of the populace, which were often castigated as "rustic" or "vulgar" and regarded as expressions of ignorance and superstition. The culture of the enlightened elite was formed by separation from—and to some extent in opposition to—the culture of the people at large.[80]

Beliefs about the weather in Britain confirm this bifurcating model to some extent, but they also demonstrate a continuing relationship between elite and popular cultures. Many among the literate elite accepted the notion of the British climate as providentially temperate, buying into the enlightened outlook that saw nature as the work of a benevolent designer. But this confidence was always liable to be disturbed by unusual or extreme weather events that refused to fit within the normal order of nature. Furthermore, those who took an interest in the weather, even if they were well connected with the institutions of the public sphere, also drew upon less respectable sources of information. As Gilbert White understood, studying the weather involved talking to a wide range of people about it. Public discourse drew upon a variety of sources in popular wisdom and inherited traditions. When people spoke about the weather, they tended to reach across the barriers between different cultural domains, for example repeating proverbs and oral lore that continued to circulate even in the most polite quarters. In this respect, the situation was at least as much one of contact and engagement between cultural traditions as of confrontation. Polite knowledge and popular culture had indeed separated to a significant degree, but there were still important exchanges between them. The rain, after all, fell on everyone, and anyone might have something useful to say about it.

We can get a sense of this process of cultural exchange by examining how the weather entered into polite conversation. In some respects, conversation was developed in eighteenth-century Britain as a cultural marker for members of the social elite. Anthony Ashley Cooper, third earl of Shaftesbury, defended its moral and epistemological value in his *Characteristicks of Men, Manners, Opinions, Times* (1711). Shaftesbury's works established a philosophical basis for the ethic of politeness; he inspired other men of letters to think of conversation as a vehicle for polite learning, as opposed to the pedantry and dullness associated with academic scholarship. Addison took up the torch in *The Spectator*, declaring his ambition to bring philosophy from libraries and colleges "to dwell in clubs and assemblies, at tea-tables, and in coffee-houses." In an essay published in 1742, David Hume wrote of the worlds of learning and conversation and of his own aspiration to mediate between the two.[81] Works of popular science were often written as transcripts of imaginary conversations, modeling the behavior of their putative readers. Women assumed a prominent role in the audience for such "conversable" works. Having been excluded from traditional academic institutions, many of them eagerly trod the path to learning through polite conversation.[82]

Beyond the works of essayists and popular writers, conversation also assumed prominence in polite society. It was the primary mode of interaction in the public sphere, a key to personal success in the institutions of sociability and commerce. It was seen as essential for navigating the new urban gathering places, especially the coffeehouses that were famous as places of unfettered discourse. And it was also indispensable in semiprivate encounters, such as mixed-sex gatherings in bourgeois homes. To succeed in the art of polite conversation, one had to be agreeable in company, show sensitivity to acquaintances or strangers, and be entertaining without giving offense. From the late seventeenth century on, a series of manuals gave advice on how to accomplish this. They claimed to teach readers how to be "complaisant" or sociable, how to avoid seeming rustic or pedantic. They cautioned that conversation was not a competitive sport, that interlocutors should not be contradicted or lectured. They urged men to seek female company for its civilizing effect, and advised women on how to converse decorously with men. They sometimes gave examples of bad conversational tactics: pedantic disquisitions on specialized subjects, for example, or graphic descriptions of one's ailments and medical treatment.[83]

Certain topics were considered too hot to handle in polite conversation. Religion and politics, in particular, were very ill-advised choices, since they risked stirring up impolite rancor or at least harming people's feelings. It was safer to talk about the weather than these things, though advice manuals generally cautioned against resort to the topic, suggesting that its banality would reflect poorly on the speaker. *The Lady's Preceptor* (1743) warned, "If the Occasion of the Visit does not afford you a Subject for Conversation, take care not to be so unprovided with one, as to be obliged to the Weather or the Hour of the Day for your Discourse."[84] For the insufficiently prepared conversationalist, however, the weather was one of the most obvious topics. It was common knowledge, about which opinions were not likely to differ too drastically; it was neither political nor sectarian. Handled correctly, it should not be pedantic or divisive. As Addison noted, it was one of the prime "Publick Topicks." The instrument maker George Adams wrote in 1790 that the fact that the weather was of immediate interest to mankind as a whole "is evident from its constantly forming a principal topic of their conversation."[85] Samuel Johnson thought it peculiar that "when two Englishmen meet, their first talk is of the weather; they are in haste to tell each other, what each must already know"; but the value of such talk as a social lubricant was clear.[86] A century later, Richard Inwards noted that "the state of the weather is . . . the usual text and starting-point for the conversation of daily life."[87]

Johnson and Inwards probably had it right. Weather discourse was often "first talk," the "starting-point" of polite conversation. It was a way of initiating an interaction, of engaging a conversational partner with whom one wanted to talk about other matters. In Samuel Richardson's *Clarissa* (1748), the news and the weather are said to be "such nonsense as Englishmen generally make their introductory topics to conversation."[88] This was what Oscar Wilde was pointing out when he had Gwendolen say, in *The Importance of Being Earnest*, "Whenever people talk to me about the weather I always feel quite certain that they mean something else." Alternatively, a remark about the weather could be a way to acknowledge an individual in public without engaging them any further, saying something that would not invite prolonged conversational exchange. Occasionally, it could be used to deflect a line of discussion or to change the subject. The weather, in other words, was an ancillary tool of conversation, a pretext for it rather than a principal subject of it. Such talk is a paradigm example of what linguists call phatic communication, in which the primary meaning lies not in what is referred to but in the social bonds consolidated by the exchange. Someone who did not realize this, who made the mistake of contradicting something that was said about the weather, or of discoursing about it at length, would be thought to have transgressed against the norms of polite conversation.[89]

There were various things one could politely say about the weather in a conversational context. Johnson asserted that "an Englishman's notice of the weather is the natural consequence of changeable skies, and uncertain seasons."[90] It seemed that another of the providential benefits of the British weather was that it gave people something to talk about. Its unpredictability meant that one could always comment on how one's fears had been quelled or hopes disappointed. Departures from what was expected at a particular season were also worthy of note. Since the British thought of their weather as perennially moderate, any episode of notable heat or cold could furnish material for a remark. By the late eighteenth century, it was also apparently acceptable to say that the weather was not what it used to be, that the character of the seasons had changed within memory.[91] Whether or not this was true was a matter of dispute among experts, but it seems to have been a commonplace comment.

In talking about the weather, the British people were able to draw upon a rich legacy of traditional sayings and proverbs. These were already beginning to be documented in print in the sixteenth century; they were compiled particularly assiduously by nineteenth-century folklorists. M. A. Denham's collection of weather proverbs of 1846 was followed by those of Charles Swainson, Richard Inwards, and others.[92] Many of the sayings were already centuries old when they were printed, some going back as far as classical antiquity. Some were specific to particular localities, like the adage about Bredon Hill echoed in the Worcestershire diary of 1703. Similar maxims about clouds on local hilltops presaging rain are recorded from many other places.[93] Most sayings were geographically unrestricted, and a few traveled across national boundaries. A large proportion of them offered short-term prognostics, signs of what the weather was about to do that could be seen in the sky or on earth. These included the appearances of the sun, moon, or clouds, which way the wind was blowing and how fast, the time of year, and the behavior of animals, birds, and insects. Maxims about how to read these signs constituted the oral lore of what was called "weather-wising." There were also many sayings about how to forecast the weather of a season up to several months ahead. The likely yield of harvests could be predicted, it was said, by noting the weather in the preceding summer, spring, or even winter. Conversely, the early ripening of harvest fruits was thought to foretell an early and snowy winter. But there were also many proverbs that were not prognostic at all. They simply commented on what the weather was doing and perhaps reassured people that it was consistent with what had happened in the past. Thus "Drought never bred dearth in England" offered encouragement that rain could not be long delayed. "When the wind's in the east, it's good for neither man nor beast" implied that people just had to put up with cold winds while they lasted. "April showers bring May flowers" buoyed up those who were awaiting the arrival of spring.

Some of these sayings are still in circulation today. Television weather forecasters still interject maxims such as "March comes in like a lion and goes out like a lamb" and "An English summer—two fine days and a thunderstorm," which have been in documented usage for centuries.[94] Folklorists and anthropologists have mapped the incidence of weather proverbs in contemporary conversation, finding, for example, how frequently they are spoken by parents to their children. A proverb, it has been said, is "an utterance that asserts itself independently of any utterer—continuously, as it were, or indeed eternally." [95] Weather proverbs carry the authority of anonymity and the comfort of contact with an apparently timeless tradition. They connect what happens on a particular occasion with the normal run of things, reassuring people that everything that occurs is consistent with the way things usually are. In this respect, proverbs reinforced the idea in eighteenth-century Britain of a providential national climate, buttressing it with the authority of traditional wisdom.

Some arbiters of politeness in the eighteenth century nonetheless saw them as inappropriate in refined conversation. As elite participation in rural customs waned and the authority of oral tradition weakened, members of the middle classes were advised to purge their speech of its proverbial baggage.[96] Satirists, however, observed that conversation, even in the most refined circles, was frequently filled with proverbs and clichés, an observation echoed by modern literary scholars. Jonathan Swift wrote a satirical *Treatise on Polite Conversation* (1738) entirely filled with the overused phrases found in elite discourse, including a number of weather sayings that originated in popular tradition. Those who published collections of proverbs agreed that it was foolish to accept them all uncritically, and it would be boorish to reel them off like Sancho Panza in *Don Quixote*. Nonetheless, traditional sayings were thought to preserve a wealth of popular wisdom about the weather that should not be overlooked. The naturalist John Ray published *A Collection of English Proverbs* in 1670, including many about the weather. Some of them, he was sure, were "superstitious and frivolous"; most would "as often miss as hit" if subjected to test; but all were worthy of preservation. A similar view was taken by the Kentish doctor Thomas Fuller, whose *Gnomologia* (1732) included more than six thousand proverbs and adages, a significant number of them concerned with the weather and the seasons.[97]

Weather proverbs were given greater prominence by the works of a number of pastoralist writers, who argued that dwellers in the countryside possessed a kind of wisdom that city folk would do well to take seriously. The anonymous *Knowledge of Things Unknown* (1743) included numerous weather maxims ascribed to husbandmen and shepherds. The compilation was a heterogeneous one, claiming the authority of ancient figures like

Pythagoras and Ptolemy alongside the timeless wisdom of the countryman. Its advice for weather forecasting was based partly on astrological methods, partly on predicting the weather of a season from that on a particular saint's day. It also described how shepherds foretold the impending weather from the behavior of livestock, birds, and bees.[98] The most widely read compilation of such maxims was *The Shepherd of Banbury's Rules,* composed by John Claridge in 1670 and republished with a commentary supposedly by John Campbell in 1744.[99] Campbell's introduction to the later edition lauded the eponymous shepherd for his natural semiotic abilities, acquired by spending a lifetime outdoors: "Every thing in Time becomes to him a Sort of Weather-Gage. The Sun, the Moon, the Stars, the Clouds, the Winds, the Mists, the Trees, the Flowers, the Herbs, and almost every animal with which he is acquainted. All these I say become to such a Person Instruments of real Knowledge."[100] As natural "instruments," the shepherd's weather signs were said to be more reliable than artificial instruments, such as the barometer, and they had the advantage of forecasting conditions for days, weeks, even months ahead.

Campbell's declared purpose in expanding and republishing Claridge's book was to make available a tradition of popular rural knowledge. It is notable, however, that he called upon two other sources of authority to support the credentials of the shepherd's maxims. First, he mentioned classical authors, "all the wisest and gravest Writers of Antiquity," who had recorded a substantial number of natural signs of the weather. The poet Virgil and the natural historian Pliny were the classical authors to whom he turned most frequently for precedents for the shepherd's sayings. But behind these Roman authors stood a lengthy tradition of Greek writings on weather signs and their relation to the seasons, going back ultimately to Hesiod's *Works and Days.* Second, Campbell tried to confirm the shepherd's findings by reference to the ideas of contemporary natural philosophers. The experimental philosophy, he claimed, "generally speaking enables us to give a fair and rational Account of almost all the Phænomena taken notice of by the Shepherd of *Banbury.*" Among his points of reference were the records of weather observers—"some of our great naturalists, who had kept Journals of the Weather for many Years"—who had established that certain winds recurred with seasonal regularity. Campbell endorsed the providential outlook that underwrote the work of the diarists and had found expression in the concept of the benevolent national climate. He insisted that all changes of the weather, even violent storms, were part of God's benevolent design: "All Weathers are at some times seasonable, which shews that they are good in themselves, and only accidentally evil."[101] His reliance on classical tradition and journals of the weather indicates Campbell's position within learned culture. His enthusiasm for the rural wisdom of the shepherd of Banbury was not the direct expression of an authentic popular tradition. Indeed, he seemed to have lifted a fair proportion of his weather maxims from John Pointer's book of a couple of decades earlier. The success of the supposed shepherd's lore attests more to a prevailing mood of pastoral nostalgia among the literate elite than to rural culture as such.

The same could be said of John Mills, an agricultural writer and fellow of the Royal Society, who published his own comments on the shepherd of Banbury's rules twenty-six years after Campbell's edition. Mills has been connected with the "georgical" movement, which drew upon classical sources of inspiration, such as Virgil's pastoral poetry, to encourage attention to agricultural improvement.[102] He was also well acquainted with the work of experimental philosophers and weather diarists; he had kept a journal for eleven years at Oundle in Northamptonshire and advised other farmers to do the same. He held up as a model the work of the Oeconomical Society of Berne, which had kept records of the weather and its effects on crop yields. To develop the utility of weather observations for agriculture, he advocated paying close attention to the accumulated wisdom of country people, represented by Claridge's compilation of weather maxims. Vernacular knowledge of this kind was not superstition, according to Mills, but a truly "natural" philosophy, revealed to rural individuals through their unmediated experience of the normal pattern of climatic events.

These pastoralist writers gave proverbial weather lore a currency and status among the social elite. They integrated native sayings with the weather signs that had come down from classical antiquity, conferring upon them an enhanced respectability and linking the worlds of popular and learned culture. They also recognized a certain coincidence of aims between rural people and systematic weather observers. Both could claim the authority of experience, though of rather different kinds. And both tended to see the natural world as governed by a benevolent providence, expressed in the seasonal recurrences of certain kinds of weather and the visible signs of its imminent changes. Given these parallels, it is not surprising that those who studied the weather often made reference to oral traditions, proverbs, and rustic lore. In 1785, Benjamin Franklin discussed the proverb "As the day lengthens, the cold strengthens," which had been catalogued by Ray a century earlier.[103] Franklin noted that it was quite true that the lengthening of the period of daylight in January did not immediately bring alleviation of winter's cold, and he explained the reason: the sun's rays at that time of the year were still too oblique to warm the air significantly.

Luke Howard was another investigator of the weather who engaged with popular lore quite systematically. Originally inspired by the summer haze of 1783, the Quaker meteorologist was encouraged through decades of studying and recording the weather by the conviction that it could ultimately be shown to reflect God's providential care of creation. Notwithstanding "perpetual fluctuations, and occasional tremendous perturbations," he wrote, "the balance of the great Machine is preserved." Howard's observations, at Plaistow in Essex and later at Tottenham, culminated in the two-volume work *The Climate of London* (1818–20), in which he proposed to rescue meteorology from "empirical mysteriousness, and the reproach of perpetual uncertainty," by diligent measurement and record-keeping.[104] Memory alone was an unreliable source of information, he explained. People were misled into thinking that the character of the seasons was changing, because their recollections of the weather a few years back were quite imperfect. In fact, according to Howard, the regular return of each season fulfilled God's promise to humanity after the biblical deluge—the promise, symbolized by the rainbow, never again to wreak such devastation. Notwithstanding the fallibility of people's memories, Howard insisted that meteorologists had much to learn from those with practical experience. Farmers and mariners, for example, "become weather-wise by tradition and experience; and are often able to communicate the results of a certain local knowledge." [105] Concerning the course of the winds, indeed, "the experience of our navigators . . . outruns science." [106] Even proverbial sayings that might appear to be pure superstition, like the well-known adage about rain on St. Swithin's Day (15 July) forecasting forty days' rain to follow, were worthy of serious discussion. Howard was keen to "do justice to popular observation" on this matter, and concluded that a showery period frequently would begin about that time, though it was unlikely that the preceding period would have been any drier in a typically wet English summer.[107] Overall, Howard's attitude to vernacular weather lore was far from dismissive; though he clearly saw scientific meteorology as something different—and a more appropriate way to pay tribute to God's goodness and wisdom—he believed it could only benefit from an openness to popular tradition.

Howard also recognized the value of published reports of unique and spectacular weather events. He complained about the lack of specificity in such reports: "The *language* of these accounts is . . . commonly vague and unphilosophical: a hard gale of wind is too often 'a tremendous hurricane,' and frost and floods, hail and thunder, are too frequently stated to have been the most severe and destructive 'in the memory of the oldest persons living!'" [108] But he found himself unable to resist the temptation to introduce such dramatic narratives into his work. In April 1807, he quoted a newspaper report from Lancashire of "the most tremendous thunder and lightning ever remembered by the oldest persons." [109] On later occasions, freezing rain and hailstorms, tornadoes, and lightning bolts found their way into his monthly summaries, accompanied by incidental details of witnesses and victims. In 1816, as England coped with another anomalous season—an unusually chilly spring and summer that retarded the growth of vegetation and severely reduced the harvest—Howard again grappled with the problem of reconciling extreme events with the general benevolence of providence. He gathered reports of unseasonable cold throughout Europe and North America and calculated that the average temperature was five degrees Fahrenheit below normal. What became known as "the year without a summer" sounded a strange echo of 1783, when Howard's enthusiasm for meteorology had first been kindled. Again, the fascination exerted by such occurrences was undeniable, notwithstanding the difficulty of reconciling them with belief in the providential benevolence of the climate. [110]

Howard was far from uncritical about popular lore and vague reports of weather wonders. He wrote that "there is no subject on which the learned and the unlearned are more ready to converse, and to hazard an opinion, than on the weather—and none on which they are more frequently mistaken." [111] He was nonetheless committed to the value of communication across the gulf between the learned and the unlearned. In contrast with William Marshall, whose attempt to assess the value of weather maxims had led him "into a labyrinth apparently endless," and who was tempted to give up the whole thing in disgust, Howard wanted to keep the lines of communication open. [112] His receptiveness to oral tradition was typical of most observers of the weather, from the Worcestershire diarist of 1703 to Richard Townley on the Isle of Man in 1789–90. Knowing the weather required listening to what local people said about it, even if their remarks were sometimes judged mistaken. Frequently, proverbial weather-wisdom confirmed the providential outlook of the elite observers, offering confirmation of an underlying divine plan. Some weather maxims could be traced back to classical writers like Aratus and Theophrastus, providing a learned pedigree for what might otherwise be taken for vulgar lore. And the interest of middle-class people in rural weather lore was also enhanced by a kind of nostalgia for the primitive that became increasingly fashionable from the middle of the century. For all these reasons, elite and popular cultures did not inhabit separate worlds when it came to the weather. Middle-class intellectuals who watched the sky realized that they also had to attend to common experience and the sayings of their unlettered neighbors.

More than a century after the Great Storm, Howard was more confident than his ancestors of the providential regularity of the British weather. His underlying faith does not seem to have been shaken by any of the unusual events he witnessed. He could draw encouragement from a century and a half of regular weather recording, which had consolidated public belief in the steady benevolence of the nation's climate, a stability that supported its agricultural and commercial activities. Daily fluctuations were part of the overall picture, a feature of the British weather that kept the inhabitants mentally alert. But anomalous seasons and truly extraordinary events were more difficult to assimilate. They inevitably drew attention from experts and from the people at large. Howard himself had been challenged to take an interest in meteorology by an occurrence of this kind. And whenever they happened, they pointed up the uncertainties of the science, its failure to predict such events or to reconcile them with the pattern of normal expectations. For this reason, anomalous weather signaled the limits of Enlightenment science. Strange weather phenomena showed the natural world in its most recalcitrant aspect, continuing to resist attempts to bring it within the pale of scientific reason.

Anomalous events also made clear how much weather observers depended on oral reports and uncertain information. One had to trust what one heard, being forcefully reminded that a science of the weather was built upon the speech of many informants. Attending to what people said about the weather, investigators found it was a mixture of traditional lore and exaggerated claims, experiential wisdom and what the enlightened called "superstition." These features were particularly exposed at times of unusual or extreme weather, but they pervaded social discourse about the climate at other times as well. Any study of the weather involved engagement with the heterogeneous things people said about it: the folklore, the proverbs, the uncertain reports, and fallible memories. The weather constituted a common domain in which elite and popular discourse intersected; it was public property, the concern of society as a whole. Thus, while scrutinizing their atmosphere, British intellectuals were also viewing a likeness of their own culture. Their consciousness of their climate reflected back to them an image of the society they inhabited: a substantially literate culture still imbued with oral practices; a partially urbanized society still rooted in the traditions of rural life; a culture that embraced science and technology while still exhibiting vestiges of magical thinking. Overall, it was a society in which the ideals of enlightenment were only incompletely realized.

Recording and Forecasting

[Dr. Johnson] again advised me to keep a journal fully and minutely, but not to mention such trifles as . . . that the weather was fair or rainy.   JAMES BOSWELL · *Life of Johnson*

Why is it that showers and even storms seem to come by chance, so that many people think it quite natural to pray for rain or fine weather, though they would consider it ridiculous to ask for an eclipse by prayer?   HENRI POINCARÉ · *Science and Method*

ALL HUMAN EXPERIENCE unfolds in the dimension of time. We shape accounts of our days, of events that happen to us, even of our lives as a whole, to make the passage of time meaningful for us. Experiences of the weather are also by their nature temporal. Weather is just what happens in the atmosphere as time passes. It is often inserted into written narratives as a counterpoint to the temporality of human lives. References to the weather can set the tone for each day's record in a daily journal, or thread through the story of a journey or a sojourn in a particular place, or punctuate the account of a public event or an individual's biography. Since people make weather meaningful by incorporating it in their own structures of temporality, changes in the means by which time is measured or recorded will change how the weather is experienced.

The eighteenth century saw significant changes in the methods for recording and measuring time. For many people in Britain, time became more comprehensively structured and more precisely measured. At the end of the century, members of the new industrial workforce found themselves subjected to the mechanical time-discipline of the factory, a pattern very different from the rhythms of agricultural life. But well before this, the spread of clocks and watches had already standardized the divisions of the day for much of the urban population, who increasingly arranged their daily lives according to quite exact measurements of hours and minutes. A growing consciousness of public time had been manifested among the urban middle classes since the late seventeenth century.[1] At another level, awareness of the standard calendar was spread through the media of newspapers and almanacs. Traditional ways of dividing the year, by the seasons or by agricultural or ecclesiastical festivals, were displaced by the uniform civic calendar. The confrontation between old and new customs was particularly intense at the time of the calendar reform in 1752. Then, the adoption by Britain and its colonies of the Gregorian calendar already used in continental Europe required a break with the traditional dating of festivals and some disruption of the seasonal patterns of rural life. The price was reckoned worth paying, at least by the educated elite, in order to gain the advantage of a uniform system of dating that united the "civilized" world. At the most general level, the "Newtonian" scale of absolute time became the framework for dating the whole of human history, which was integrated with the timing of such astronomical events as eclipses and the orbits of comets. The scale was extended backward to the earliest events of secular and biblical history and forward as far as one cared to extrapolate the motions of astronomical bodies.[2]

These developments had an important bearing on experiences of the weather. In fact, one might say that "the weather" as we understand it—as a quotidian occurrence—was constituted through regular record-keeping governed by the clock and calendar. Before this, people thought of "climate" as the environmental conditions affecting life in a particular place, and of "meteors" as extraordinary appearances in the air, but they didn't think of the weather as something that was happening at every hour of every day. Michel Serres has claimed that the time of "meteors" is distinct from—and historically prior to—the uniform timescale of "weather." In his view, adoption of a homogeneous scale of time measurement tended to normalize atmospheric phenomena, reducing them to the regular and the repeatable. In other words, phenomena of the air ceased to be part of an understanding of time as *kairos* (a discontinuous set of significant sacred events) and became elements in a (continuous, secular) *chronos*. They were

noticed not only when they burst upon the scene as apparently preternatural or portentous occurrences, but all the time. The weather assumed the
form that is familiar today, being understood as whatever is happening in
the atmosphere at the specified points of a uniform scale of time. The table
of days was supplied by the civic calendar, and clocks were used to time
observations by the hour. Those who set out to compile weather diaries
took this given temporal framework, initially empty but homogeneous,
and went on to fill in the gaps in the table they had drawn up. Something
had to be noted in every time slot; even the absence of wind or precipitation
at a particular moment constituted a data point.[3]

The change Serres identifies did not occur overnight; it unfolded as
part of a prolonged cultural transformation, rather than a sudden shift of
philosophical perspective. The weather was normalized by recording it on
a uniform scale of time, which came into use as part of the cultural changes
of the eighteenth-century Enlightenment. Reduced to its underlying regularities, the British weather was identified with the island's climate, which,
as we have seen, was regarded as God's gift to the nation. As we have also
seen, extraordinary meteorological phenomena did not stop occurring, but
systematic recording forged a new understanding of the weather in which
they played a less prominent role. Most weather diarists conceived of their
activity as part of a public program of natural history, a contribution to the
collective enterprise of building up knowledge of nature. By compiling a
body of records, they hoped eventually to be able to predict the weather,
as laws of astronomy had been found to govern the motions of planets and
comets. Adoption of a uniform framework of time allowed observers to feel
that they were participating in a communal activity directed toward this
goal, although the actual derivation of laws of the weather was postponed
into the indefinite future. The prospect of limitless progress in the growth
of knowledge was itself a feature of the new view of history that saw it as
structured by a homogeneous scale of time.

The weather diarists were prepared to wait until laws of the weather
could be extracted from their records. They were disciplined men and wom-
en, who took the long view of the importance of their task. Some of them
were sustained by a sense of duty inherited from traditions of Christian
spirituality, which had long found expression in the practices of journal-
writing. Patience was counseled by the Edgiock diarist in 1703, who wrote
that "to be oracular in weather asks no small pains & labour, & can alone
be founded on a vast & extensive science."[4] But the population at large
clearly felt the need for more immediately useful knowledge. Oracles were
demanded, notwithstanding the caution of the record-keepers. The indefi-

nitely postponed fulfillment of the diarists' project left a gap, which was filled by a range of schemes for weather prophecy. As we have seen, traditional methods of "weather-wising"—which interpreted such "signs" as cloud formations, appearances of the sun and moon, and the behavior of domestic animals—continued to be published and to circulate in conversation throughout the eighteenth century. New instruments like the barometer were generally interpreted in relation to these traditional prognostic signs, as we shall see in the next chapter. And there were also traditional methods for predicting the weather for a coming season, which continued to feature in popular almanacs. Some of these were based on astrological techniques; others used the weather on a particular day to forecast the character of the approaching season. Neither procedure had any legitimacy from the point of view of the weather diarists, but in the absence of any more authoritative means of forecasting, methods of seasonal prognostication continued to be widely used. The persistence of these methods was a further sign of the limited diffusion of enlightened attitudes in society at large, a symptom of the survival among the general population of more deeply rooted notions of time.

The Discipline of the Diary

What motivated someone to keep a weather diary? The question is worth asking because the chore has generally been recognized as a tiresome one. Some authors wrote journals that summarized the weather over a season or a month at a time, but many compiled their records on a daily basis. Some read instruments—especially the thermometer and barometer—once, twice, or even three times a day, usually at more or less specified hours. The task demanded meticulous exactitude and remarkably steady habits. The observer had to be dedicated to the routine and to live an even, uninterrupted life, usually residing in the same place for an extended period of years. Ralph Thoresby, a Yorkshire virtuoso and compiler of an extensive personal journal, considered adding a weather record in the 1690s, "but bethought myself of the tediousness of it." When Clifton Wintringham brought out his journal of more than a decade's observations of weather and diseases in York, in 1727, he admitted to "the continual tediousness in making these observations over several years." As the twentieth-century meteorologist Gordon Manley put it, "Prolonged maintenance of daily observations demands an odd and uncommon type of enthusiasm." [5]

Aside from rare medieval precedents, that odd enthusiasm first took hold of people in the late seventeenth century. As we have seen, the Royal Society began at that time to provide encouragement for the work of weather diarists, and the goal of printed publication gave them a sense of participation in a public enterprise. But even before this, the first weather records had formed part of private journals compiled for rather more personal reasons.[6] The Essex clergyman Ralph Josselin, who kept a diary at Earls Colne from 1644 until just before his death in 1683, frequently noted the weather in the course of his journal. His interest in it seems to have been twofold. Weather conditions had a substantial impact on agricultural production, directly affecting how much food was available to Josselin's family and their general level of prosperity; and the weather was seen as an indicator of God's favor or disfavor toward the diarist, his family, and the community around him. Josselin did not use instruments to make measurements, nor did he record conditions every day, but he kept an eye on storms and the general character of the seasons. He tended to note bad weather more often than good, painting a vivid picture of how vulnerable rural life was to the capricious climate and how directly people felt themselves subject to God's judgments.[7] The weather also featured in the diary of Andrew Hay, a Presbyterian landed gentleman from Lanarkshire in Scotland, the surviving portion of whose record covers the period from May 1659 to January 1660. Hay detailed his activities on each day, his readings and the sermons he heard, and ended each annotation with two short lines, one giving a summary statement of the state of his soul or mood, and the other a brief description of the weather. The juxtaposition of the two elements is often striking, although the author made no explicit links between them. Statements such as "This day was full of temptations and sad" or "I found this day my heart hard to gather" were set alongside comments like "A most fearfull, constant rain all day" or "Windie in the morning and raine afternoon." Rather like the Edgiock diarist four decades later, Hay seems to have been noting parallels or contrasts between the state of the air and his inner feelings.[8]

In diaries like these, the weather featured in a general system of providential accountancy that frequently motivated the compilers of Protestant spiritual journals.[9] God was thought to be rewarding or testing the faithful through sending good or bad weather; he might deliver judgments or admonishments through destructive storms. Since the Reformation, British Protestants had largely dispensed with the rituals that Catholics had used to try to avert God's wrath. Ringing church bells during storms or praying to particular saints to relieve drought had been dismissed as popish superstitions.[10] Lacking recourse to magical techniques of this kind, Protestants nervously scrutinized the weather as a sign of their moral standing in God's eyes. They sought reassurance from external phenomena that they were destined for salvation rather than damnation. The absence of certainty on this most vital matter was a source of perennial anxiety. In his *Anatomy of Melancholy* (1621), Robert Burton coined the term "religious melancholy" to describe mental pathologies connected with religion, including people's despair about their prospects of salvation. The melancholy humor lying at the root of such conditions was labeled by Burton *balneum diaboli*, "the devil's bath." And one cause of such melancholy was disorder in the atmosphere: "The devil many times takes his opportunity of such storms, and when the humours by the air be stirred, he goes in with them, exagitates our spirits, and vexeth our souls."[11] Later in the seventeenth century, religious delusions came to be classed as a species of "enthusiasm" and were traced to causes such as excessive susceptibility to the qualities of the air. For devout individuals liable to depression, this raised the urgent issue of the source of such feelings. There could be no more important question than whether one was miserable because one's soul was in peril or simply because of the weather.[12]

As the Puritan belief in particular providential interventions declined in the course of the seventeenth century, explicit comments on how God was acting through the weather tended to become less frequent. Samuel Pepys, writing his diary in the 1660s, recorded unusual storms, periods that were especially hot or dry, and early or late frosts, but he never seems to have reflected that God was sending any kind of message by these means. He noted the weather on particular public occasions, like Charles II's coronation, and what people said about it, but he held himself aloof from those who claimed to interpret its providential significance. Many diarists continued to monitor the weather in connection with agricultural concerns, sometimes recording the prices of crops at market on the same pages. In other documents, records of the weather accompanied notes on the diarist's state of health. Many seventeenth-century diarists used their journals to keep track of their bodily ailments and the treatments they were undergoing, and sometimes comments on the weather went along with these notes, though direct connections were rarely made.[13] Individuals who felt themselves subject to melancholy had a reason to seek correlations with the weather. The experimental philosopher Robert Hooke sometimes noted occasions when a melancholy mood descended on him. His diary, begun in 1672 and continued to 1693, recorded his social transactions with members of the scientific community, inventors, and tradesmen, along with his illnesses and the medications with which he dosed himself. Initially, Hooke's diary also included a weather record, maintained on a daily basis from March 1672 to April 1673.[14] Although Hooke was not someone who manifested particular religious piety, his practice of noting melancholy feelings in conjunction with weather observations echoed earlier practices of Protestant spirituality.

Hooke sustained a dual focus in his diary: on external circumstances, on the one hand, and on the internal conditions of the observer's body and mind, on the other. The technique of the daily journal could encompass both realms. Hooke scrutinized the workings of medicines on his own body as meticulously as he made any observation of natural history; and he also understood his internal state as bearing upon his capacity for external observation. Like many of his contemporaries, Hooke believed that philosophical objectivity required suppression of the passions. The passions were believed to originate in the body; if unrestrained, they had the potential to cloud the mind. The reliable observer had to be a master of self-possession, immune from the emotional disturbances that could disrupt the reasoning powers or the working of the senses. Various techniques of regimen were put forward to accomplish this, including recommendations for diet, exercise, daily routine, and abstinence from luxurious habits, rich food, or sexual activity. Many seventeenth-century philosophers followed such rules of regimen, in the belief that accurate observation and reasoning required a form of "care of the self." [15] Hooke's concern in his diary with his bodily constitution and passions was in line with this widespread preoccupation. He was using his diary as a tool for monitoring his capacity for scientific work.

When Hooke contributed his "Method for Making a History of the Weather" to Sprat's *History of the Royal Society* in 1667, he was appealing for others to tread the path of disciplined observation. The document was a programmatic statement of the Royal Society's aim of encouraging widespread weather observation, a project supported by Robert Boyle, John Locke, and others. The intention was to produce some degree of uniformity in the reports the society wanted to collect. To this end, instructions were given for tabulating daily observations in eight categories: winds, temperature, humidity, pressure, clouds, prevailing illnesses, incidents of thunder and lightning, and the tides. Observations for a whole month should be presented on a single sheet of paper, correctly arranged so that they could be taken in at a glance. This was said to be necessary for the process of induction as Francis Bacon had described it, or "requisite for the raising *Axioms*, whereby the Cause or Laws of Weather may be found out." In view of the difficulties that stood in the way of realizing this aim, it is rather striking how much was *not* specifically stipulated in Hooke's instructions.

Little was said about the need to standardize and calibrate instruments, for example. And as to the qualifications of the prospective observers, the document stated only that they should be "some one, that is always conversant in or neer the same place." [16] Aside from immobility, no other personal qualities were specified.

Hooke's instructions, and the various initiatives that followed, probably spurred some individuals to undertake weather observations, though nobody seems to have submitted a record of exactly the kind he wanted. Some diarists were no doubt sustained through their inherently tedious task by the sense that they were contributing to the public stock of knowledge, mapping the national climate as part of a collective enterprise. Nonetheless, for most weather observers in the eighteenth century, the motivation must have been largely personal. Compiling a weather diary was not part of anybody's job; the duty was a self-imposed one, even though it was sanctioned by social institutions and the public sphere. The degree of discipline it required came from within. This suggests that weather diaries might be related to the earlier tradition of spiritual journals, expressions of the same kind of care of the self that Hooke displayed in his own diary. Notwithstanding their apparently quite different focus — upon external events in the physical environment rather than upon the internal state of the diarist's soul — weather journals could be seen as products of a certain kind of exercise of self-formation.

The late seventeenth and early eighteenth centuries saw the flourishing of a variety of texts that allowed the writer to develop and express a sense of self, including personal journals and autobiographies. These have plausibly been seen as secular progeny of Protestant spiritual journals and more distant descendants of classical and Renaissance models. [17] Diaries of the weather have not usually been classed with these works, since their outward orientation gave them a very different subject matter. Unlike the Worcestershire diary of 1703, most weather journals suppressed all manifestations of personal subjectivity, although (again, unlike that diary) they usually identified their authors by name. In fact, diarists came to be recognized as reliable observers of the weather to the extent that they concealed their feelings and personal qualities. They manifested their objectivity as observers by making themselves, in effect, disappear. Like others who recorded natural phenomena, they rendered themselves transparent in order to warrant the accuracy of their account of the world around them. [18] Personal idiosyncrasies had to be hidden so as not to impede the translation of private records into the common stock of knowledge. Weather journals were therefore a paradoxical vehicle for self-development — one in which self-effacement and concealment of personal motives were norms of the genre.

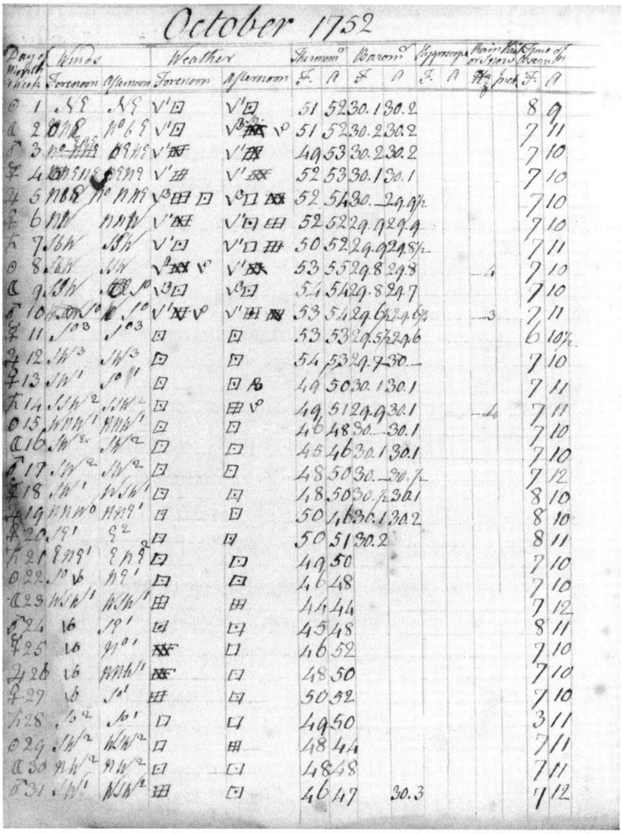

FIGURE 7 · Meteorological observations by John Murray.
Murray, a ship's surgeon, compiled a fairly typical
weather journal. This page records general conditions,
readings of thermometer and barometer, and the
time of observation for the month of October 1752.
Courtesy of Wellcome Library, London.

One weather diarist who explicitly avowed a spiritual motive for his work was the Quaker physician John Rutty (1698–1775). Rutty worked as a medical practitioner in the city of Dublin for several decades after his arrival there in 1724; he also became a leading member of the small Quaker community in Ireland. The focus of his scientific studies was on environmental causes of health and disease. He published on mineral waters and the natural history of the Dublin region; his *Chronological History of the Weather and Seasons and of the Prevailing Diseases in Dublin* (1770) drew upon a weather journal he kept from 1725 to 1766. He also wrote religious works, expressing his interest in the Christian devotional tradition and his personal commitment to methods of spiritual discipline. After his death, his *Spiritual Diary* (1776) was edited and published by a fellow Quaker. The techniques he used to compile a weather journal and those he applied to spiritual self-development have intriguing similarities. For more than forty years, Rutty was making daily annotations of his observations of the weather and the diseases he observed in the course of his medical practice. He was also maintaining an intimate record of his spiritual achievements and lapses, castigating himself daily for moral failings visible to no one but himself. The importance of bodily discipline and temperance emerges strongly in both spiritual and medico-meteorological records. Rutty saw any overindulgence in food or drink, or any degree of luxuriousness in habits of life, as a threat to religious and physical health alike. His own asceticism was legendary among his contemporaries; rumor had it that he sometimes dined on nothing but nettles.[19] He regularly noted even slight excesses in his consumption of food or drink. As the editor of his spiritual diary put it, he tended to "take blame to himself, where he thought that he had passed the line of rectitude, though even but a hair's breadth." [20] By rigorous temperance, Rutty strove to preserve his body as a vehicle both for spiritual development and for acute observation of the natural world. Both tasks required a strenuous mortification of the flesh in order to suppress disturbances of the passions.

Rutty suggested various ways in which his weather journal could be justified in religious terms. Its aim was to reveal the regular workings of divine providence in nature, to address what he called "a common species of blasphemy[:] . . . crying out against the weather." In addition, he was convinced that the daily contemplation of human suffering from disease and natural disasters was morally salutary. Approaching the end of his decades of weather recording, Rutty spent a "sweet evening in a review" of his work. He reflected on the length of life God had given him to complete it, and gave thanks. And he simultaneously reviewed his spiritual state, "even the history of my progress and regress, and restoration through the Divine Bounty," concluding, "Surely, here has been no small industry in nature: Lord, supply the defects in grace!" In some respects, however, Rutty worried that his scientific interests were a distraction from his religious preoccupations. He repeatedly promised to renounce natural inquiries and dedicate himself solely to spiritual concerns. He also blamed himself for succumbing to the allure of worldly glory as a published author, taking a perverse consolation in July 1759 from "a mortifying, but wholesome repulse on application to the Royal Society." [21] Clearly, the moral calculus surrounding scientific pursuits was finely balanced. Rutty was so anxious about the dangers of all kinds of indulgence that he suspected himself of enjoying even painstaking observational work too much. Nonetheless, the self-discipline he applied to his scientific inquiries, and particularly to his methodical journal-keeping, was complementary to the spiritual improvement that was his overriding concern. Both required repeated and meticulous observation, as he explained in the preface to his spiritual diary, and both demanded humility and self-abnegation by the observer.

Modern psychologists could no doubt offer diagnoses of Rutty's mental condition on the basis of his spiritual diary. The editor of the text suggested he had a "hereditary choler in his natural temperament," which he turned inward against his own passions.[22] Rutty himself mentioned incidents of "doggedness," perhaps indicative of melancholy or depression, which brought him low at times and kept him in bed. Some of his contemporaries had very little sympathy. Samuel Johnson greeted the publication of the diary with glee, laughing heartily over a review that punctured the author's earnest self-importance. Rutty's spiritual seriousness made him an object of ridicule to many people, though James Boswell reflected that the register of his state of mind, while "frequently laughable enough, was not more so than the history of many men would be, if recorded with equal fairness."[23] Rutty might have been an eccentric individual, but Boswell at least recognized the fundamental honesty of his attempt to chart the progress of his own soul. His record of the struggle to control his body and his passions manifested a basic sincerity and determination to tell the truth. These virtues also underwrote the authority of his weather record.

Notwithstanding his eccentricities, Rutty's discipline and self-denial are echoed in the work of other weather observers. Other diarists shared his asceticism, his scrupulous exactitude, and what one biographer called his "mind trained to appreciate the importance of little things," even if they did not maintain an interest in the tradition of spiritual self-discipline.[24] The personality of Thomas Barker (1722–1809), squire of Lyndon Hall in the county of Rutland and a long-term weather observer, was similar in many respects.[25] Barker's life was characterized by the immobility that Hooke had considered desirable; he was born and died in the same house, and rarely spent any time away from it in the course of his eighty-seven years. He recorded his first weather observation at the age of eleven, began keeping a systematic journal when he was fourteen, and kept it up for nearly sixty years. During the 1770s, 1780s, and 1790s, he published abstracts of his annual registers of the weather in the *Philosophical Transactions*, records that modern meteorologists have judged among the most reliable of the period. Twice a day, month after month, year after year, Barker read his thermometer and barometer, at times that he measured to the minute by his clock. He collected rainfall in his rain gauge and analyzed the measurements to tease out patterns of variation through the seasons. He also occasionally reported unusual weather phenomena he had observed: a waterspout in 1749, for example, and a solar halo in 1761.[26] Many of these interests were shared with his correspondents Gilbert and Henry White in Hampshire, with whom Barker exchanged weather records and letters on natural history.

At the time of his marriage in 1751, Barker was said by a friend to be "naturally prone" to "extreme Abstractedness & Speculativeness." Perhaps it is a sign of those qualities that he registered the nuptial occasion only as the cause of an interruption in his weather journal. On previous visits to Gilbert White's home, he had recorded "at Selborne"; when he married White's sister Anne, the note read "at Selborne, etc."[27] Barker's asceticism was also remarked upon; an almost lifelong vegetarian, he was said to be still thin and athletic in his sixties. An interest in long-term observation of very slow natural processes seems to have complemented his lengthy program of weather recording. He weighed a piece of flint and buried it under the soil of his garden in 1757, digging it up twelve years later to weigh it again. The experiment was designed to determine if the stone would increase in size over time due to mineral accretions. Barker concluded that it had in fact diminished, but speculated that the result might have been different in chalky soil. He measured the rate at which his fingernails grew, by painting marks at the roots and waiting until they grew out. He recorded having done this no fewer than four times in the 1750s.[28] From a certain point of view, these are likely to seem trivial activities, symptoms of an almost pathological concentration on minute incidentals. However, they are also tributes to monumental patience and a kind of expansive vision of the dimension of time in which natural processes unfold. They could be said to look forward to Charles Darwin's observations, a century later, of the long-term effects of very slow processes like the building up of coral reefs or the circulation of soil by earthworms.[29]

Barker's records and journals were entirely concerned with external observable facts, not with his internal or spiritual state. He occasionally reflected on the religious implications of natural disasters, suggesting that cattle diseases and swarms of locusts were divine punishments that should call people to repentance. But in general, he shared with Rutty a faith in the providential regularity of the laws of nature. His interest in the orbits of comets, a preoccupation shared with his grandfather William Whiston,

FIGURE 8 · "Farmer G——e Studying the Wind &
Weather" (1771). An anonymous political satire,
with George III as a weather observer who ignores
the disorder in his own household. Courtesy
of The Lewis Walpole Library, Yale University.

testified to the hope that prolonged observation would reveal underlying
natural laws, an aspiration for meteorology as much as for astronomy.[30]
To accomplish this would require diligent habits of record-keeping, sustained over very long periods of time. Barker qualified himself for the task
by monumental patience and immobility, meticulous attention to precise details, abstraction from much of the social world around him, and a
routine of ascetic regimen. In these respects, his regime of self-discipline

constituted a kind of "care of the self," although he did not himself draw out its connections with the traditions of Christian spirituality. He did, however, exemplify the emerging modern construct of scientific objectivity, in which the emotions and the demands of the body are suppressed as a precondition of reliable scientific observation.

Other weather observers seem to have shared the same ethos. Although family members and servants no doubt made possible the diarists' concentration on their task, they were never mentioned in the records.[31] William Derham noted the occasion of his marriage in rather the same way that Barker did, apologizing for a gap in his journal caused "by my absence from my observations about necessary affairs."[32] In general, the diarists mentioned personal circumstances or their own state of health only to account for lapses in their records. Thus, Clifton Wintringham noted in the summer and autumn of 1725, "I myself about this time was seized by a serious and lengthy disease, which kept me from these observations for four months."[33] In the last three decades of the century, Thomas Hughes, a physician at Stroud in Gloucestershire, mentioned his own confinement due to rheumatism and lumbago at times when it prevented him from making observations of the weather and diseases in his locality.[34] At such rare moments, the body of the observer asserted itself as an obstacle to the routine recording of observations. In the normal run of things, however, the weather diarists might have been taken for entirely disembodied observers. Unlike the Edgiock diarist, they did not present themselves as subject to the weather's influence, but as detached and objective witnesses of it. Indeed, their stoical perseverance without regard to their own bodily state is indicated by the fact that some authors are known to have kept up their diaries until just a few days before their deaths.[35] Like other individuals who aspired to the objectivity demanded of modern scientific practitioners, they represented themselves as passionless beings, masters of self-control.

Becoming a weather diarist was a peculiarly self-denying form of self-expression. The authors of these works effaced signs of personal subjectivity in order to establish their status as objective observers. To compile documents that would be acceptable contributions to the public scientific enterprise, they suppressed their own bodily reactions to the weather, even as they explored its effects on the health of others. It seems telling that such impersonal texts were often written by authors who were noted for their self-denial and ascetic practices. They adapted methods of self-discipline that had originally served spiritual aims, in order to transcend the limitations of their own corporeal state. In their weather records, they represented themselves as free of the encumbrances of the body, as detached and passionless observers of the natural world. The disciplinary practices by which this was achieved were counterparts to the social and institutional structures within which the diarists worked. Practices of care of the self complemented the social mechanisms for regulating the passage of time, for example. Together, they enabled the diarists to sustain themselves in their prolonged task and to merge it into the collective project of accumulating knowledge of the weather.

The Calendar and the Seasons

In agricultural societies, people develop knowledge of the weather in connection with various measures of time. They look ahead to the day to come, judging its likely character by the appearance of the sky in the morning or the sunset the evening before. They wonder whether a dry spell will continue long enough to allow the harvest to be gathered, or how many days will pass before a drought is broken. Perhaps the most important temporal scale for gauging the weather is derived from the seasons. Awareness of the seasons has been found to be fairly universal in human cultures, even among people who live in tropical regions where the weather scarcely varies at different times of year. In temperate climates, the growth of crops is limited to certain months, so seasonal weather is of critical importance to agricultural productivity. Farming communities generate a large fund of lore about what weather is to be expected at what times and how to predict what a coming season will be like. This lore is integrated with the rhythms of rural life, the festivals and other communal activities that punctuate the cycle of the year.[36]

In Western literature, seasonal weather lore can be traced back to the *Works and Days*, attributed to the Greek poet Hesiod and dated to the eighth century BCE. The poem traced the cycle of the seasons by reference to astronomical signs like the rising and setting of specific stars. It gave instructions to farmers as to what they should be doing at each period to take advantage of the weather they could expect. As Laura Slatkin has noted, in Hesiod's poem, "that which is 'timely' . . . becomes a figure both for the ordered life and for a standard of appropriateness within it."[37] The text stood at the head of a lengthy tradition of classical Greek works, the *parapēgmata*, which provided astronomical calendars and listed the weather likely at each point in the year. These in turn gave rise to stone tablets with holes drilled in them, in which pegs could be inserted and moved from hole to hole as each day passed. Set up in public places, these calendar stones reinforced a sense of the timeliness of particular conditions and the importance of understanding what was seasonable at a particular date. They complemented social rituals in their function, transferring the timely order of the heavens to the arrangements of human collective life.[38]

In early-modern Europe, almanacs took over some of the functions of the ancient *parapēgmata*. Cheap and mass-produced, these short printed tracts were astonishingly widely distributed among the semiliterate mass of the population. They printed seasonal weather lore in the form of proverbs and fragments of verse alongside astrological predictions derived from the motions of the planets. Much of this lore was also preserved in oral culture and in the customs regularly performed during the year; it emphasized the importance of seasonal rituals to keep in step with the sequence of natural events. Weather that departed from seasonal expectations posed obvious problems for agriculture and was taken as a sign that things were generally in disarray—a bad omen for human affairs.[39] As well as conveying broadly what conditions were to be expected at particular times, vernacular lore presented specific rules for predicting the character of the coming seasons. Some of these rules relied on indications from animals and plants. If squirrels were seen to be gathering large stores of nuts in the autumn, for example, or apples were found to have particularly thick skins at harvest, one should take warning of a hard winter to come. Many rules were tied to dates in the calendar, usually identified with the feasts of particular saints. The weather occurring on those days was said to be especially significant, giving a sign of what was to come in the ensuing days, weeks, or months. There were a number of these "prognostic days" in the calendar, and belief in them seems to have survived throughout the eighteenth century; many proverbs concerning them were still current when they were collected by nineteenth-century folklorists.[40]

The days at the beginning of the year were often thought to set the pattern for all that followed. The weather on the twelve days of Christmas was scrutinized as a key to what could be expected in the twelve months of the coming year. The first day of January was also ascribed prognostic importance, although the new year did not legally begin on 1 January in England until 1752. Other days regarded as similarly prophetic included 25 January (St. Paul's Day), 1 July, 2 November (All Souls' Day), and 25 December (Christmas Day).[41] As is still well known, rain on St. Swithin's Day (15 July) was thought to foretell rain for forty days thereafter. According to the memorable doggerel included in an almanac of 1675, "If St. Swithin weeps, the proverb says, / The weather will be foul for forty days."[42] Some of the beliefs concerning Candlemas Day (2 February) survive to the present in the customs associated with Groundhog Day in the United States. In many European countries, the weather on that day was taken as a predictor of how soon spring would arrive. Animals such as bears and badgers were studied for what their behavior revealed, just as the groundhog in Punxsutawney, Pennsylvania, is quizzed as to whether he can see his shadow.[43] A similar significance was ascribed to St. Bartholomew's Day (24 August), the weather on which was thought to foretell that of the autumn as a whole.

Most of the sayings concerning prognostic days tied them to the temporal framework of the ecclesiastical calendar. They shared this feature with agricultural festivals and customs, which by the eighteenth century had long been fixed according to the dates used by the church. This meant that some prognostic days always corresponded to a certain date in the civic calendar, while others moved around with the changing dates of Lent, Easter, and Whitsuntide. Country people were willing to accept that the feasts connected with Easter were moveable ones, and they apparently accepted shifts in some of the days used for seasonal prognosis with the same equanimity.[44] Popular belief even countenanced "borrowed days," when winter conditions more typical of March returned unexpectedly in early April. On the days that March borrowed from April, it was reported, Scottish people would not lend or borrow household articles, lest they be used for witchcraft against their owners.[45] In Ireland, these days were associated with Bó Riabhach, the legendary brindled cow, who is supposed to have complained about the harshness of the March weather.[46]

Seasonal weather prognostics were rooted in a heterogeneous collection of calendar customs and lore. They were symptomatic of a flexible attitude to temporal measurement, a way of handling time that was integrated with the seasonal routines of rural life and adaptable to the agricultural circumstances of particular locales. Use of saints' names to label the days provided a kind of formalization of traditional dating practices, but it did not obliterate local variations or smooth out the different degrees of significance attached to different periods of the year. In this respect, as Keith Thomas has noted, "nothing did more than the ecclesiastical calendar to reinforce the conviction that time was uneven in quality."[47] It was precisely this unevenness and variability that the uniform civic calendar was intended to overcome. In the course of the eighteenth century, the civic calendar's authority was asserted against surviving rural and local practices. Various proposals for reform were floated in the first half of the century, justified in the language of enlightenment and scientific rationalization. A parliamentary statute legislated for the introduction of the Gregorian system in 1752, mandating that Britain and its colonies switch dating practices to match those of continental Europe by dropping eleven days between the second and the fourteenth of September that year. There was a significant degree of reluctance to abandon the traditional Julian calendar, which had tied customary events like rural festivals and fairs to the cycle of the seasons and the harvesting of crops. The dates of some festivals were shifted back in the calendar, so that they would keep the same position in the agricultural year. In many rural locations, the Old Style dates continued to be used, notwithstanding the reform, because they were familiarly associated with certain events. Although it seems that stories of widespread riots—with protesters complaining that they had been deprived of eleven days of their lives—were mythical, there is no doubt that the calendar reform was tacitly resisted in some quarters.[48]

Those who systematically collected weather records adopted the uniform timescale of the civic calendar, aligning themselves with the program of reform. Their registers placed events within a chronological structure that was established in advance and could be extended indefinitely. The availability of this public framework made the composition of a diary more or less a matter of routine—a practice that demanded a degree of discipline and perseverance, but where the formal outlines of the writing were laid down ahead of time. Weather recorders often established the outline of the calendar graphically by drawing lines on the blank page to mark spaces for each day in the month to come. Printed outlines were available to be used by weather diarists from the late seventeenth century; they offered a series of blank spaces for each day in the month or the year. A daily record could be compiled by filling in the blanks on such a printed form with readings of instruments or comments on the weather of that day.[49] The method made a certain formal structure available to guide observations; it contributed to the reconstruction of the weather as a quotidian phenomenon that occurred at all points on a continuous scale of time.

The notion that particular days were predictive of the weather to come was not easily reconciled with such an attitude to time. Generally, enlightened commentators on the weather were scornful of the popular belief in prognostic days. John Ray, the naturalist who collected weather proverbs in the late seventeenth century, pronounced, "I think all observations about particular days superstitious and frivolous."[50] The agricultural writer John Mills thought that fundamentally sound knowledge about the pattern of the seasons had been attached by "monks and designing priests" to specific saints' days in order to consolidate the power of the medieval church.[51] Even when elite authors began to take an interest in popular weather lore, they were generally condescending about prognostic days. In the late nineteenth century, Richard Inwards wrote that when it came to beliefs about particular saints' days, "we are constantly treading on the fringes of the veil of superstition."[52] Not all commentators were quite so disdainful, however. Luke Howard was willing to examine beliefs about St. Swithin's Day sympathetically, notwithstanding his Quaker distaste for Catholic tradition.[53] So was the Catholic writer Thomas Forster, who sought in the early nineteenth century to recuperate popular ideas about the seasons and their place in traditional religious observances. In his *Perennial Calendar and Companion to the Almanack* (1824), Forster claimed that the belief in the significance of certain days as weather indicators would be "found not to be devoid of Truth." Seasonal changes, as yet poorly understood, did tend to occur at particular dates, and it was not unlikely that the character of a coming season could be discerned by the weather at that time. Forster admitted that the calendar reform had sown confusion in the minds of those who tried to apply prognostic days. But, he went on, "the popular Belief in the Rules outlived the Change of Style, and the Husbandman and the Astrologer still consult the critical Days as heretofore."[54]

Whether scornful, condescending, or sympathetic to popular weather lore, elite writers all reflected an awareness that a gulf had opened up between vernacular and enlightened beliefs on this subject. Disagreement about the plausibility of prognostic days was a sign of more basic differences in how time itself was understood. The arguments over calendar reform were themselves symptomatic of deepening cultural divisions, as urban development and the spread of enlightened attitudes drew middle-class people away from their connections with the agricultural practices of the countryside. As the historian Maureen Perkins has written, the general trend was one in which "time became disconnected from communal memory, from the planets, from the cycles of nature. It was now to be just the march of numbers, neutral, unchangeable, inescapable."[55]

The uniform scale of numbers that constituted the public calendar could be expanded indefinitely into the future and the past. In relation to the future, it replaced the apocalyptic expectations previously invested in portentous meteorological events with a confident anticipation of unending progress. Eschatology gave way to the prospect of an open expanse of time stretching away forever.[56] The practice of compiling chronological weather records was consistent with a vision of natural inquiry as a collective public enterprise that would continue accumulating knowledge for a very long time. Weather journals were necessarily limited in geographical and temporal range, so it could always be argued that more data were needed from more observers before conclusions could be drawn. The physician and weather recorder William Hillary wrote in 1740, "There must be many Collections, and of much longer Continuation, obtained with the greatest Exactness, before we can draw such Aphorisms as are certain and conclusive from them." [57]

In relation to the past, the uniform scale of time allowed for the application of Newtonian "absolute time" to the understanding of history.[58] The *General Chronological History of the Air* (1749), by the Sheffield physician and statistician Thomas Short, went all the way back to the biblical flood, which the author dated to anno mundi 1657. Beginning at that point, he embarked on a lengthy catalogue of plagues, floods, pestilences, earthquakes, famines, and other extreme meteorological and epidemic events. Occurrences that Christians and Jews had perceived as miraculous interventions were included in a single series with the marvels recorded by pagan historians. All were reduced to episodes in a uniform chronological sequence. Volume one of the two-volume work took 494 pages to reach 1717; volume two picked up at 1711 and included in a seamless continuum the records of observers such as Wintringham (in York) and John Huxham (in Plymouth). Short stitched the records of these contemporary weather observers into a continuous temporal fabric that extended back to the time of the biblical patriarchs. His work extrapolated the public calendar to construct a chronological framework for meteorological phenomena even of the distant past. The approach tended to normalize even the most extraordinary weather events by placing them in a regular table of dates.[59]

Recording the weather on a uniform scale of time did not imply that the seasons were disregarded. The idea of prognostic days was scorned by most of the weather diarists, but they understood very well that each season was distinguished by its characteristic weather. In fact, the program of systematic recording suggested new ways in which these seasonal patterns might be studied and methods of prediction perhaps devised. The seasons themselves were widely regarded as an aspect of God's providential design of the cosmos. Newton himself thought it plausible that God had tilted the earth's axis in relation to the plane of the ecliptic in order to cause them. The great Swedish naturalist Carolus Linnaeus introduced the genre of the seasonal calendar: a catalogue of the plants that flourished at each time of the year. The Scandinavian example was adapted to cover British plants in *The Calendar of Flora* (1761), by the botanist Benjamin Stillingfleet. A more popular version for children was *The Calendar of Nature* (1784), by the Dissenting physician John Aikin. The general message of these works was that the cycle of the seasons was part of God's plan, that each kind of weather was suitable to the time of year in which it occurred, and that plants and animals were crafted by providence to cope with the cyclical changes.[60]

A number of those who compiled weather diaries claimed to have identified seasonal regularities that had not previously been appreciated. John
Rutty in Dublin and Gilbert White in Hampshire insisted that experienced observation would show the seasonality of even what seemed extreme or abnormal weather conditions. John Campbell told readers of his
edition of *The Shepherd of Banbury's Rules* (1744) that "some of our great
naturalists, who had kept Journals of the Weather for many Years, have
found that the same Wind blows every Year very near the same number
of Days, and that there is a regular Continuance of different Winds annually in every Country." Campbell also referred to Hooke and Derham as
authorities for the conclusion that wet and dry years, and hot and cold ones,
balanced one another out if one took a sufficiently long view of the record.
Systematic observation had demonstrated a "balance of the weather which
Providence has established," confirmation of the scriptural assurances
that God had settled things "by weight and measure." As Campbell saw
it, scientific observers of the weather had uncovered a new dimension of
God's providence—its manifestation over the course of prolonged periods
of time. By recording the weather systematically, one could deepen one's
admiration for the divine handiwork, building upon the rustic appreciation
of the cycle of the seasons a more refined understanding of "the constant
and settled order established by the will of that Almighty Being, which
order we ordinarily call Nature."[61]

Campbell acknowledged that townsfolk had lost the intimate connection with nature enjoyed by people like the shepherd of Banbury. A gap
had opened up between elite and popular experiences of the weather in
relation to time. But he was nonetheless trying to rehabilitate an aspect of
rural tradition—the notion of seasonality. An improved appreciation of
the seasonal qualities of weather would follow from the collection of more
observations. Then one might come to grasp not merely that the seasons recurred in a regular way, but that a more profound balance lay behind what
seemed like chaotic irregularities from day to day. This realization of the
depth of providential benevolence was something for which the weather
diarists could claim credit; following their work, it became a commonplace in writings on natural theology. William Whewell, Master of Trinity
College, Cambridge, whose book *Astronomy and General Physics* (1833)
was one in the series sponsored by a legacy from the Earl of Bridgewa-
ter, devoted a chapter to the seasons as proofs of God's design and another
chapter to the constancy of climates. Seasonal alterations of weather and
annual changes in vegetation were providentially adapted to one another,
Whewell argued. And although the succession of weather conditions in the

short term might appear entirely irregular, it is "easy to see, with a little attention, that there is a certain degree of constancy in the average weather and seasons of each place." The same degree of attention that revealed the underlying regularity of the British climate could also, according to Whewell, bring to light its systematic differences from the climates of other parts of the world.[62] These differences were those on which the British had long congratulated themselves, considering them a divine blessing on their nation. Whewell was adding to them another reason for gratitude to the deity: that systematic recording had also uncovered deeper regularities in the patterns of the weather, going beyond the simple and obvious fact of the recurrence of the seasons.

In this way, systematic recording of the weather was said to have built upon the universal recognition of the pattern of the seasons by showing that more profound regularities underlay apparent variations. The timeliness emphasized by seasonal weather lore since Hesiod was given a new depth of meaning. Prognostic days were relegated to the domain of vulgar error, but elite observers reinterpreted the notion of seasonality in terms of their own providential conception of the order of nature. They insisted that the appropriateness of particular weather conditions could be proved by statistical analysis of long-term records. In this way, deeper levels of temporal order were revealed by disciplined inquiry operating within the framework of the public calendar. Popular knowledge was to be superseded but not entirely rejected. The traditional sense of the timeliness of weather patterns was thought to have been confirmed by enlightened scientific research. Stripped of its errors, vernacular lore was shown to have a core of truth.

Forecasting by the Heavens

Order, wrote Sigmund Freud, is "a kind of compulsion to repeat," the function of which is to spare oneself hesitation and indecision. The compilers of weather diaries strove to bring order to the phenomena of the atmosphere by compulsive repetition of their gestures of looking, reading, and recording. They probably succeeded in bringing a measure of order to their own lives thereby, and perhaps did something to assuage the anxiety of passing time by adopting the reassuring practice of repeated ritual. Freud also noted that "man's observation of the great astronomical regularities not only furnished him with a model for introducing order into his life, but gave him the first points of departure for doing so."[63] As we have seen, the order of the heavenly bodies was indeed the source of the regularity imputed to the seasons in the first weather calendars. The public display of the ancient calendar stones, their pegs moved daily from hole to hole by some authorized individual, shows how the astronomical order was used as a basis for arranging human affairs.

Turning to the heavens for a key to the weather, the ancients went beyond simply using the movements of the sun and stars to mark the passing seasons. They studied celestial phenomena and the motions of the planets in the hope that they could be used for prediction. The moon was reckoned to have a significant influence on rainfall: its position, its phases, and the inclination of its axis to the horizon were all thought to be meaningful. The planets, comets, and eclipses were also believed to shape the weather. Each planet was said to impinge upon certain elemental qualities of the atmosphere, and its effects would be modified by the zodiacal sign it occupied. Particularly important were the "aspects" of the planets—their angular separation from one another in the sky—whether in conjunction (close together), in opposition (on opposite sides of the sky), or at some other significant angle. Ancient astronomers, culminating in Claudius Ptolemy in the second century, formalized the methods of astrological weather forecasting to take all these factors into account. They acknowledged that astrological influences would have different effects in different places according to the prevailing qualities of the air, thereby dovetailing their knowledge with the ancient discourse of local climates. Astrologers needed specific local expertise in addition to their general knowledge in order to forecast the weather at a particular time in a particular place.[64]

Ancient techniques of astrological weather forecasting were kept alive in Europe during the Middle Ages and benefited from a general revival of astrology during the Renaissance. In the early-modern period, astrology was developed as a demanding field of expertise. It attained a height of popularity in England in the middle of the seventeenth century, when leading practitioners earned fortunes from their businesses and millions of copies of astrological almanacs flooded from the presses. Much of the enthusiasm for astrology at this time was focused on what it might foretell about political and military events, but the prospect of predicting the weather was also of interest.[65] In 1671, the Scottish astrologer William Cock laid out the procedures of astrological weather forecasting in his *Meteorologia*. Cock suggested that twelve separate factors had to be weighed in giving a forecast, including the natures of the individual planets, their positions on the zodiac, their aspects, the season, the latitude and situation of a country, and the sign of the zodiac associated with it. Astrological weather forecasting was a complex task, in which a judgment had to be made of the respective importance of many different factors. Cock echoed the ancient insistence that local knowledge was part of the skilled astrologer's qualifications. For example, a spell of rain might be foretold by the conjunction of Mars and Venus in the constellation of the Crab; but the rain would fall more heavily in Scotland than in England, because the former kingdom was subject to the sign of the Crab, while the latter was not.[66]

Astrological weather forecasting had a significant place in the "reform" movement in astrology, which tried to establish a new foundation for the art after the Restoration in 1660. The would-be reformers, who included Joshua Childrey, John Gadbury, and John Goad, turned their backs on political and religious factionalism, which had compromised astrologers of the previous era. They emphasized weather forecasting as an example of what their art could genuinely hope to accomplish. Astral influences on the weather were entirely natural, they argued, and could legitimately be studied without the taint of conjuring or superstition. The three all had substantial connections with the Royal Society, though none was elected a fellow, and they invoked the authority of the society's prophet, Francis Bacon, who himself had called for a reformed astrology that would forecast the weather.[67] Gadbury published an annual almanac with the Greek title *ΕΦΗΜΕΡΙΣ* (*Ephēmeris*), in which he laid out astrological predictions for the year to come. He claimed to have drawn upon two decades' worth of records to prove the influence on the weather of heavenly bodies, especially the moon. Goad wrote a systematic defense of natural astrology in his *Astro-Meteorologica* (1686). He suggested that it would in fact be superstitious to *deny* the evident bearing of celestial events on the weather merely because they could not be fully explained. He shared with natural philosophers the conviction that the universe manifested God's benevolent design in its underlying regularities. For Goad, astrological indications were comparable to the weather signs used by husbandmen and sailors; but astrology had the advantage over "*Vulgar prognostics*" of looking ahead to a whole season rather than just a few hours. Goad claimed he had been compiling observations for three decades prior to the publication of his book, which would make him the first person known to have been keeping a systematic weather record in seventeenth-century England.[68]

The advocates of astrological reform succeeded in gaining the attention of some of the leading natural philosophers of the Restoration period. As we shall see in the next chapter, both groups shared a fascination with the barometer as a putative prognostic device. The Astronomer Royal, John Flamsteed, wrote in 1678 that Goad's predictions "come much nearer truth than any I have hitherto met with."[69] When the Oxford virtuoso Robert Plot published a diary of the weather for 1684 in the *Philosophical Trans-*

actions, he expressed admiration for the accomplishments of "the Learned Dr. *Goad* of *London,*" and called for further research on the planets' effects on weather.[70] Robert Boyle's *General History of the Air,* edited by John Locke and published after the death of the author in 1692, included a letter written decades before to Samuel Hartlib, in which Boyle indicated a strong interest in research on natural astrology. He admitted the possibility of planetary influences on the weather, even while acknowledging the impostures of paganism and superstition. Lunatics, melancholics, and epileptics were all known to be affected by the celestial bodies, noted Boyle, and it seemed likely that all of the planets had some influence on atmospheric composition by drawing forth effluvia from the earth. What was needed to resolve the matter was systematic record-keeping, "an History or Diary of the Observations of the Weather, and its Changes in all Respects, and then an Account of the several Places, Motions, or Aspects, each Day, of the several Bodies of the Heavens."[71] Apparently, one motive for Boyle's advocacy of weather recording was to test the claims of astrology.

By the beginning of the eighteenth century, the interest in astrology among natural philosophers had declined rather dramatically. Some Restoration virtuosi had taken a much less sympathetic view than Boyle, and their scorn was echoed in the new century by literary authors such as Joseph Addison and Jonathan Swift. In a surprisingly short period, the cultural elite turned against the ancient art, condemning it as a relic of the superstitious past. But printed almanacs continued to be very popular, even though the number of titles was winnowed down and the volume of sales declined from its seventeenth-century peak. At the end of the eighteenth century, there were still six nationally circulated titles that contained significant astrological content, including the most successful of them all, *Moore's Almanack* or the *Vox Stellarum,* which was selling more than 350,000 copies annually.[72] Weather forecasts also continued to be featured in some almanacs throughout the century, occupying a single column in the table devoted to each month of the year. Reading down the column, one would encounter a more or less continuous narrative of the predicted weather, which one could match approximately to the days of the month listed in an adjacent column. Usually, the same column contained symbols for the phases of the moon, eclipses, and significant aspects of the planets. These were the astronomical events from which the weather forecasts were supposedly derived, though the derivation was never spelled out. If one took the trouble to compare different almanacs' forecasts for the same month, one would find that the same astronomical data yielded somewhat different predictions of the weather. In any case, the predictions were notoriously vague, both as to the likely conditions and how long they would last. Here, for example, is the forecast of the renowned astrologer John Partridge for the month of November 1685: "The Months [*sic*] beg[ins] with vio[lent] weather and holds so for some days. Cold and wet, and so holds for some days. Warmer and more pleasant but soon alters, & grows stormy and wet with cold. Moist, but very moderate, and a remission of cold also. Cold and sleety and holds." [73]

A skeptical reader might conclude that the elaborate apparatus of the astrologer's calculations had yielded little more than what one might arrive at by pure guesswork. November in Britain is very likely to be chilly and wet. Readers perhaps did not expect astrological forecasts to be much more precise than common expectations for the season, and insofar as stipulations were given of the timing of weather conditions, they were understood to be frequently unreliable. Of course, whether one thought that this uncertainty refuted astrology as a whole depended on one's attitude to the art. The astrologers themselves took refuge in the complexity of their calculations and the need to apply judgment to balance the different factors. They acknowledged the many possible sources of error as a way of defending their skill. And even the fact that skepticism about astrology had become widespread among the educated in the eighteenth century did not mean that such people would have no interest in astrological weather forecasts. Since classical antiquity, skepticism had surrounded the predictions of oracles and the results of divination, but the fascination with them survived, just as it does among readers of newspaper horoscopes today. For this reason, the extent of underlying *belief* in astrology is not really the most important factor in accounting for its continued appeal. Even skeptics might be interested in reading the weather forecasts in almanacs and working out to what degree their ambiguous prophecies were fulfilled and to what degree they were not.

Many almanacs seem to have been used in just this kind of way, checking predictions against the conditions that actually occurred. Surviving copies frequently bear written annotations, recording events as they happened, including the weather. Users scribbled these notes in the margins of the pages or made use of space deliberately left blank by the printers. [74] Used in this manner, almanacs provided an outline for weather recording within the framework of the calendar, rather like the printed grids employed for compiling weather diaries throughout the century. Other publications incorporated records of the weather in the year gone by, in place of—or as a supplement to—the forecast for the year to come. The *Ladies' Diary*, an almanac that also included letters, poems, and mathematical puzzles,

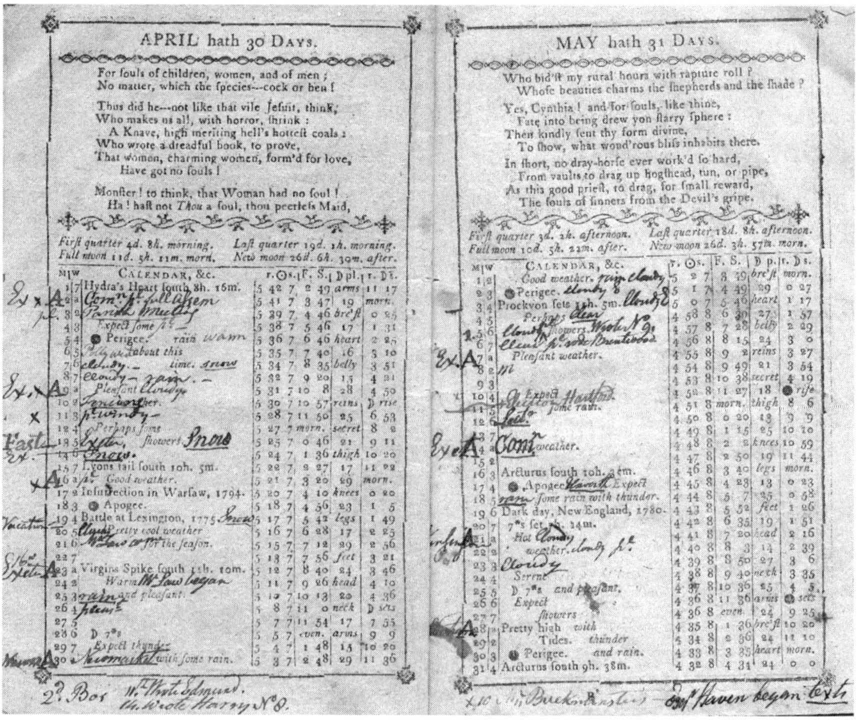
APRIL hath 30 DAYS.

For souls of children, women, and of men;
No matter, which the species---cock or hen!

Thus did he---not like that vile Jesuit, think,
Who makes us all, with horror, shrink:
A Knave, high meriting hell's hottest coals:
Who wrote a dreadful book, to prove,
That women, charming women, form'd for love,
Have got no souls!

Monster! to think, that Woman had no soul!
Ha! hast not Thou a soul, thou peerless Maid,

First quarter 4d. 8h. morning. Last quarter 19d. 1h. morning.
Full moon 11d. 5h. 11m. morn. New moon 26d. 6h. 30m. after.

MAY hath 31 DAYS.

Who bid'st my rural hours with rapture roll?
Whose beauties charms the shepherds and the shade?

Yes, Cynthia! and for souls, like thine,
Fate into being drew yon starry sphere:
Then kindly sent thy form divine,
To show, what wond'rous bliss inhabits there.
In short, no dray-horse ever work'd so hard,
From vaults to drag up hogshead, tun, or pipe,
As this good priest, to drag, for small reward,
The souls of sinners from the Devil's gripe.

First quarter 3d. 1h. afternoon. Last quarter 18d. 8h. afternoon.
Full moon 10d. 5h. 22m. after. New moon 26d. 3h. 57m. morn.

FIGURE 9 · Annotated pages from an almanac. Pages for April and May from a copy of *The New-Hampshire Diary: Or Almanack for the Year of our Lord 1797* (Exeter, NH: D. Ranlet, 1796). Alongside the printed weather predictions, the owner has recorded the conditions that actually occurred, together with personal appointments. Courtesy of Special Collections, Dimond Library, University of New Hampshire.

eschewed astrological weather forecasts from its inception in 1704. The engineer Henry Beighton, after he assumed the editorship in 1714, began to incorporate his own weather record, leading to quips that the *Ladies' Diary* had decided to play it safe by predicting only events that had already taken place.[75] In 1778, Henry Andrews, a teacher of mathematics at Royston in Hertfordshire, brought out *A Royal Almanack and Meteorological Diary*, in which he supplemented the traditional table of planetary aspects and weather forecasts with his own diary of the weather from the previous year. Presumably, the reader was meant to use both sets of data conjointly to assess what was likely to happen in the year ahead.[76]

These adaptations of the genre suggest that almanacs—like diaries and periodicals—directed their readers' attention to the quotidian quality of the weather, its unfolding within the uniform framework of the civic calendar. From time to time, extraordinary or violent events would interrupt this smooth continuum. Almanacs would forecast some violent storms in the course of the coming year, usually hedging their bets as to their exact timing, and readers often inserted written notes when they occurred. Clearly, almanac users understood that the predicted dates for such events were approximate at best. Insofar as they offered predictions more specific than seasonal generalities, the forecasts of the almanacs were widely acknowledged to be uncertain. But the anomalous character of extreme events was itself testimony to the normal equanimity of the climate, its general coincidence with seasonal patterns. In this sense, the almanacs confirmed the overall timeliness of the British weather, demonstrating to all their readers that one could be more sure of the repetition of seasonal patterns than of their sporadic interruption by storms and other anomalies. The texts conferred upon the weather at least this measure of the regularity of celestial phenomena, whatever readers believed about the planets' actual influence on the atmosphere.

There was one respect in which the question of the influences of the heavenly bodies remained open, even among those who otherwise scorned astrology. Many commentators clung to the conviction that the moon had a significant bearing on the weather. Opinion among the enlightened elite was skeptical of seasonal prognostics, and planetary aspects were no longer taken seriously by astronomers; but the moon obviously influenced the ocean tides, and it seemed entirely plausible that it exerted a similar force on the atmosphere. Throughout the eighteenth century and beyond, people debated how this might shape the weather. The apparent coincidence of women's menstrual cycles and lunatics' mood swings with the moon's phases suggested that at least some individuals were attuned to these effects. In the 1680s Goad asserted, as if it were common knowledge, that the moon was responsible for aches and pains, the mysteries of generation and conception, the motions of bodily humors, the behavior of animals, and the growth of plants. A century later, Allen Hall, a clerk from Derbyshire, and Benjamin Hutchinson, a vicar in Huntingdonshire, both said they had embarked on their programs of weather recording in order to determine the moon's influence. From their observations, they reached opposite conclusions, Hall finding support for a series of popular prognostic rules that tied changes in the weather to the moon's phases, while Hutchinson concluded that "the opinion long received of such an influence" could not be confirmed.[77]

In the last years of the century, Luke Howard decided that the problem was still worthy of serious examination. He was struck at first by what seemed like a correspondence between the level of the barometer and the

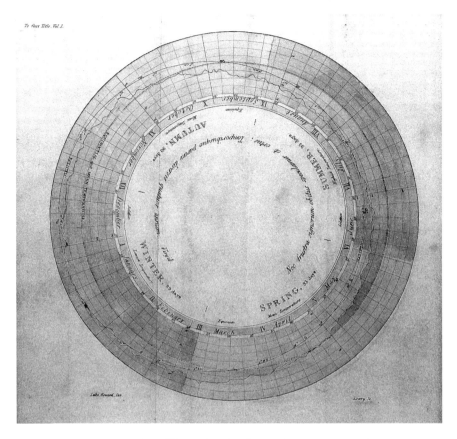

FIGURE 10 · Luke Howard's cycle of annual temperature variation.
As part of his inquiry into cyclical changes of weather, Howard prepared this
diagram of the variation in temperature throughout the year. The scheme
shows how temperature varies above and below the overall average and in relation
to the equinoxes and solstices. Courtesy of Wellcome Library, London.

moon's phases: the mercury generally seemed to stand lower when the moon
was new or full, and higher at the intervening quarters. More extended
observation called this generalization into question, but Howard was not
willing to give up the search for some kind of link. He analyzed the Royal
Society's weather record over a ten-year period and detected evidence for
a connection between lunar phases and changes in the weather of the kind
that "has long served to direct the predictions of the Almanac-makers." [78]
Having charted the seasonal variations in the barometer for a number of
years, he felt he could factor these out of his record to allow the connection with lunar phases to emerge more clearly. This he did during the year

1806–7. He now reached a rather different conclusion from before, namely, that the barometer level is elevated at the approach of the full moon, with the apex about two days before. Then the level drops at the approach of the new moon, with a return to the mean level at the time of the new moon itself. Having established these apparent effects, Howard went on to consider how the moon's phases influence temperature and rainfall, and to expand the range of causal factors to include the altitude of the moon's path in the sky. The statistical analysis became increasingly complicated, and the effect he sought remained somewhat elusive. But Howard continued the inquiry, devoting two decades to trying to prove the existence of an eighteen-year cycle in the pattern of the seasons that would approximately correspond to the repeat period for the combined lunar and solar cycles.[79]

Howard persisted in his statistical analysis, confident that a pattern was there to be disclosed as part of God's providential design. The complicated nature of the moon's orbital motions, and the interference of the much greater power of the sun's heating and gravitational effects, made the subject especially intricate. Furthermore, the very legitimacy of this area of investigation was tainted by association with astrology. Other investigators, including Howard's predecessor and fellow Quaker, Rutty, believed that the whole subject of lunar influences on the weather should be condemned because of its link with pagan superstition. Howard admitted that meteorology as a whole was slighted by many people, who "bestow on it no more regard than the supercilious notice they would take of the Weather column in a Moore's Almanac."[80] But it was this attitude that was itself irrational, in his view. To assume that there could be no truth underlying the ancient astrological art was in fact to surrender to superstition. Howard remained committed to the notion that there was some natural connection between the heavenly motions and the apparent turmoil of the atmosphere, and that diligent observation and calculation could bring it to light.

The taint of astrology continued to surround discussion of lunar influences on the weather—and other proposed methods of forecasting—through the middle of the nineteenth century. In the 1860s, some astrologers reenacted the reform initiatives of two centuries before, attempting to raise the credibility of their art by connecting it to methods of weather forecasting.[81] At the same time, meteorologists who shared an interest in prediction fought to dissociate their programs from popular astrological writings. While laying claim to government patronage and professional status, meteorological forecasters were obliged to engage in an undignified competition for public esteem against the more widely known astrologers.

They struggled to escape from the designation "weather prophets," with its disabling suggestion of vulgar divination.[82]

The association between weather forecasting and astrology was unavoidable even in the nineteenth century because it was historically very deeply rooted, going back to the ancient calendars that forged a connection between the weather and the motions of the heavens. Recording weather within the framework of the calendar served to establish its seasonality, its timeliness in accordance with the yearly motions of the sun and stars. Though eighteenth-century weather recorders did not always venture to make predictions, the aspiration to predictability was inherent in their enterprise. Analogies with astronomy were frequently made, recalling the success of Newton and his followers in predicting the orbital paths of planets and comets. A similar program of systematic observation was expected to yield predictive knowledge of the weather, with great potential benefits. As the Irish chemist and meteorologist Richard Kirwan noted in 1787, "The motions of the planets must have appeared as perplexed and intricate [as the weather] to those who first contemplated them; yet by persevering industry, they are now known to the utmost precision."[83] The ambition to forecast the weather took its inspiration from the charted motions of the heavens, even among those who denied that atmospheric conditions had celestial causes. They used the calendar as the medium through which the weather was to be subjected to the temporal order of the heavens, and in doing so they could not entirely evade their astrological inheritance.

This was another way in which study of the weather in eighteenth-century Britain made investigators aware of the limits of enlightenment. Systematic recording, governed by rigorous self-discipline, was trying to establish knowledge that would be secure from the encroachments of ignorance. But at the same time, the enterprise itself was liable to confusion with astrology insofar as it nursed pretensions to forecast the future. The legacy of the ancient art persisted in connection with attempts at weather forecasting. Indeed, enlightened investigators themselves could not be certain what might be worth salvaging from astrological tradition. The question of the moon's influence was one topic that raised this quandary particularly acutely. The adoption by weather observers of the public framework of the civic calendar had carried with it vestiges of practices that descended ultimately from the priests of ancient Mesopotamia. For enlightened gentlemen of the eighteenth century, the inheritance was troubling, but apparently inescapable.

Barometers
of Enlightenment

For the rising in the BAROMETER is not effected by pressure but
    by sympathy . . .
For it cannot be separated from the creature with which it is inti-
    mately & eternally connected . . .
For where it is stinted of air there it will adhere together & stretch
    on the reverse . . .
For it works by balancing according to the hold of the spirit . . .
For QUICK-SILVER is spiritual and so is the AIR to all intents and
    purposes.

CHRISTOPHER SMART · *"Jubilate Agno"*

It may with truth be asserted, that it is to this instrument [the ba-
rometer] that modern philosophy owes its existence, and modern
Europe its superiority over all other regions.

RICHARD KIRWAN · *"An Essay on the Variations of the Barometer"*

ONE OF THE MOST OBVIOUS ways in which Britain changed during
the eighteenth century was in terms of its material culture. New levels
of economic prosperity brought new fashions in clothing and household
goods within reach of the affluent portion of the population. Continuing
creativity in the visual and plastic arts contributed to a reshaping of the
material world, and new technology started to change people's experiences
of work and leisure. The transformation of material culture affected the
period's fascination with the weather and climate through the widespread
adoption of new instruments. In the course of the eighteenth century,
thousands of barometers, thermometers, and hygrometers were set up in
prosperous homes, while wind and rain gauges were placed outside. The
novel artifacts captivated the attention of their owners and demanded interpretation of their meanings. They had originated among the leading

experimental philosophers in the mid-seventeenth century and, within a few decades, became common in the homes of the social elite; they were the most prevalent tokens of the new science to enter the domestic sphere. By the use of these homely artifacts—sometimes described as "philosophical furniture"—the weather was literally brought indoors, domesticated as part of the quotidian routine of many households.

Meteorological instruments entered the bourgeois home through the activities of experimental philosophers, specialist instrument makers, and those who purchased the latter's products. Thermometers and barometers became the most widely circulated accessories of "polite science," more popular than the microscopes and telescopes, air pumps, globes, and orreries (mechanical models of the solar system) also found in middle-class homes.[1] Their spread reflected the growth of a market for luxury consumer goods in the period. As objects of conspicuous consumption, they conveyed some of the values associated with the Enlightenment. Purchasers were acquiring recognized symbols of enlightened culture, which is not to say that they were all being enlisted in rigorous scientific inquiry. Possession of meteorological apparatus did not, by itself, convey any training in the practices of experimental research. Especially toward the end of the eighteenth century, precision measurement in meteorology attained a degree of exactitude among the leading specialists that most users of weather instruments had neither the ability nor the inclination to match. The devices in use among non-experts were incapable of such accuracy; they were used less rigorously and interpreted more flexibly, sometimes in ways that specialist experimenters disapproved of.

In general, when instruments are widely distributed, they tend to be used for purposes not intended by those who originally developed them—for displays of status rather than precise measurement, for example. Instruments, like other artifacts, are made meaningful in specific contexts, by becoming embedded in the practices of their users. People interpret them with reference to their own assumptions and by incorporating them into their own actions. They use artificial devices as complements to their bodily abilities, extending their capacity to experience and act upon the world. As has been noted in a recent study of the role of apparatus in making scientific knowledge, "instruments and body techniques work together."[2] That this is so was recognized in the eighteenth century, when users of scientific artifacts regarded them as continuous with the experiences of the human senses. Instruments were seen as extensions or supplements to sensory perception—not as replacements for it—since the senses were believed to be the source of all reliable knowledge. Scientific artifacts were judged by comparing them with the body's inherent capabilities.[5] Weather instruments were frequently interpreted by their users in this way. Their indications were compared with sensory perceptions of temperature or moisture, for example. They were also taken to show atmospheric conditions that impinged upon the body at a level below conscious awareness, for example changes in air pressure or aerial effluvia that could affect people's health or mood. The human body itself was sometimes thought of as a kind of instrument, because of the unconscious environmental susceptibilities that the use of the new devices had revealed.

Among weather instruments, the barometer exerted a special fascination, because it appeared to be able to predict changes before they happened. Its first investigators in the seventeenth century hoped to find laws that would allow it to be used for forecasting, at least in the short term, with a high degree of reliability. By the early eighteenth century, however, such hopes had been disappointed, and experimenters satisfied themselves by saying that the barometer's predictions were to be understood as probable rather than certain. Lists of "rules" circulated that purported to specify how the barometer could be used to anticipate the weather to come. The instrument was understood to yield signs, to be read in conjunction with other signs, such as those preserved in the traditional lore of weather-wising. This pattern of use made the apparatus into a kind of "oracle," a source of forecasts that were understood as inherently ambiguous and demanding subtle contextual interpretation. The barometer was generally seen as only dubiously trustworthy; it was regarded as liable to impose upon the gullible. This made barometers somewhat equivocal tokens of enlightenment, as they reminded people of older patterns of weather prediction by omens and lore. On the one hand, they were readily assimilated by a wide community of users as a "philosophical" means of weather forecasting; on the other, they were thought to be invested by some with an excessive degree of faith that looked rather like superstition.

The Genealogy of Weather Instruments

All scientific instruments have their origin in experiments. But to become an instrument, a piece of apparatus has to cease to be experimental. It has to acquire a stable physical form, suitable for transportation and reproduction, and become accepted as a medium for representing a specific natural phenomenon. Only once it has been accepted, or "black-boxed," can it be regarded as a tool for experimental investigation rather than an object of study. The process of becoming taken for granted is a social one, requiring a new device to be incorporated in routine practices and a consensus to be formed as to its meaning. Thermometers and barometers were developed as instruments among experimental philosophers in the course of the seventeenth century. Their designs were standardized, at least as regards their basic features, and their functions—representing respectively the temperature and the pressure of the air—were recognized. Disagreements continued to surround their applications in the different sciences, and specialized refinements were made in their designs, but the basic identities of the two instruments were stabilized within the community of specialist users.[4]

In the eighteenth century, however, thermometers and barometers came to be more widely used than any other kind of scientific apparatus. Everyone was interested in the weather, so the availability of these new tools for investigating it was seized upon by many people. In this large and diverse population of users, interpretations of the instruments continued to vary. Different opinions were expressed as to exactly what changes in the air they registered, and how. Both instruments were widely viewed as analogous to the human body in its responses to the qualities of the air. Poets, artists, and novelists played creatively on the association between the behavior of these implements and the changing moods and passions of the human frame. Caricaturists, such as William Hogarth, showed thermometers and barometers as registers of the emotional climate. Laurence Sterne mused on the possibility that the human heart might be opened for inspection like the glass tube of a barometer.[5] And the poet John Phelps used the barometer as a model of the body in its intimate connection with the physical environment on the one hand and the mind on the other:

> The pois'd *Barometer* will sink or rise,
> In Mode proportion'd to the changing Skies,
> The Air serene th'inclosed Mercury shows;
> And, as by Weight impell'd, it upward goes;
> But when dilated Vapours crowd the Air,
> Its sinking State will straitway make appear;
> Solid and fluid Parts our Frame compose,
> The Fluid through the denser Solids flows.
> Th'incumbent Air is Circulation's Spring,
> And changes various as its Weight will bring;
> The Air serene, from Clouds and Vapours clear,
> Not burnt with Heat, nor chill'd with Cold severe;
> Adjusts the Motion of the circling Blood,
> The Pulse beats right, the Circulation's good;
> Vapours and Storms aerial Weight abate,

Our Blood runs low, and languid is our State,
If Cold or Heat prevail to great Excess,
More than we ought, we then perspire or less,
Our passive Body Alterations finds,
And with our Bodies sympathize our Minds.[6]

These responses should not be marginalized as merely poetic excesses, irrelevant to the history of weather instruments as such. Rather, the rich metaphorical meanings of the thermometer and the barometer were essential to their social acceptance. These meanings were rooted in the common ancestor of the two devices, the "weather glass," known since the sixteenth century. First described by the Neapolitan encyclopedist of natural magic, Giambattista della Porta, the weather glass was already being sold to the public by London merchants in the 1630s. The simplest design, described in the early seventeenth century by the English mystic and virtuoso Robert Fludd, was a glass bulb with a long, tubelike neck.[7] The bulb was heated to dilate the enclosed air and then inverted, with the end of the tube immersed in a basin of water. The water rose up the tube, trapping the air in the bulb as it cooled. The apparatus was mounted in a stand to support the bulb, and the level of water in the tube was read as a sign of the state of the atmosphere. An alternative design took the form of a glass vessel containing air and water, with a tube attached to the lower part of the vessel and pointed upward. The upper end of the tube was open to the surrounding air. The tube would partly fill with water, its level rising and falling as the air trapped within the vessel expanded and contracted. Again, the level of water in the tube was taken to signify the state of the atmosphere in general and to foretell dramatic changes like approaching storms. An advertisement for the apparatus printed in 1631 promised that "by diligent Observacion you may foretell frost, snow or foul wether."[8]

Fludd emphasized the weather glass's significance as a kind of microcosmic symbol of the universal macrocosm and as a model for the human body. The scale on the device could be marked with geographical indicators, two "hemispheres" separated by an "equator." Then, as the water rose, it would show "that the disposition of the Ayre is by so many degrees more of Northern or Boreall nature."[9] The scale could also be marked with the names of bodily humors thought to correspond to particular temperatures of the air: bilious temperaments were identified with hot air; sanguine, phlegmatic, and melancholy ones with cold. Individuals with a particular humoral complexion could then determine whether the quality of the air was likely to accentuate their predisposition. In these ways, the apparatus

FIGURE 11 · Seventeenth-century weather glass.
A typical design of the period. The level of
water in the open spout responds to changes in air
temperature and pressure. Courtesy of Science
and Society Picture Library, London.

was understood as a "key to two worlds"—the cosmos and the body—its
motions reflecting the changes in the corresponding entities. Fludd was
not the only person to view the device in this way. In the 1660s, the Catholic
astrologer Thomas Willsford wrote that the weather glass displayed the
changing qualities of the atmosphere, "made visible by a sympathetical
imitation of the parts here inclos'd." It presented to view, he wrote, "the
state of the *Air*, whether Dropsical or Feverish, Hot or Cold . . . [and] from
thence you may visibly presage the approaching weather." [10]

By the time Willsford was writing, the leading experimental philosophers were inclined to dismiss the common weather glass as a misleading
artifact, since it responded to the effects of both atmospheric temperature
and pressure without differentiating between them. The air trapped in the
glass by the water expanded or contracted largely in response to changes

in ambient temperature, but the weather glass's ability to give a sign of approaching storms was probably due to a drop in the surrounding air pressure. Two fellows of the Royal Society, Henry Oldenburg and John Beale, wrote in 1666 that "the open Weather-glass is known to signifie nothing at certainty, having a double obedience to two Masters, sometimes to the *Weight of the Air*, sometimes to *Heat*." [11] The point was fundamental to the discrimination of barometers from thermometers, which was becoming recognized by experts at this time. The year before, Robert Boyle, drawing on several years' experience with both instruments, had condemned the traditional weather glasses as "dead engines." [12] Nonetheless, the term *weather glass* continued in general use throughout the eighteenth century, usually applied to the barometer. [13] At least some of the ideas voiced by Fludd lived on. Users continued to regard the barometer as a model of the body, and sometimes as a kind of metonym for the world as a whole. They sometimes still thought of the mercury in the instrument as having "sympathy" with the air, as the eccentric poet Christopher Smart did in the 1760s. [14] And they continued to regard the barometer as a sort of oracle, from which changes in the weather might be forecast.

Thermometers and barometers began to be distinguished from one another in experiments of the 1640s. In 1644, Evangelista Torricelli first performed his experiment of suspending a column of mercury in a tube, its open end immersed in a vessel of the same liquid open to the air. Argument subsequently swirled around the experiment, concerning whether there was a true vacuum at the sealed end of the tube and what caused the suspension of the mercury column. Around the same time, Florentine experimenters made the first devices in which a sealed tube of liquid — rather than a sample of air, as in the traditional weather glass — was used to register the temperature of the atmosphere. The 1650s saw experimentation on both devices by investigators in many European countries. The readings of thermometers were compared with bodily sensations of heat and cold, exploring the discrepancies between what the apparatus showed and what the body felt. There was debate as to whether the instruments revealed the fallibility of the senses or, rather, the senses showed the instruments to be fallacious. Barometers were taken to the top of the Puy de Dôme mountain in France and to the bottom of English coal mines, to detect changes in the weight, or "spring," of the air. In an experiment in Oxford in the late 1650s, a mercury column was placed in the receiver of an air pump and found to fall as the pump was worked, registering the reduction in atmospheric pressure as air was sucked out of the vessel. [15]

It was only in the early 1660s, however, that the barometer became an instrument for studying the weather. In pneumatic experiments prior to that, no attention had been paid to the effect on the mercury of different kinds of weather, except implicitly when it was noted that placid air provided the best conditions. A shift in the evidential context occurred among the members of the early Royal Society as they began to focus on the small variation in the height of the mercury column that would occur even if the apparatus was left in the same place, a variation that was apparently independent of heat or cold. The coining of the words *baroscope*, by Robert Hooke in June 1664, and *barometer*, by Boyle in 1665, signaled the emergence of an instrument intended to measure this variation.[16] At first, investigators explored a range of possible causes of the mercury's oscillation, searching for connections with the motions of the moon, eclipses, sunspots, and earthquakes. But changes in the weather quickly emerged as the most likely cause. At a meeting of the society on 9 September 1663, Boyle asked for mercury tubes to be kept at Oxford and at Gresham College in London, in order to observe how the variation in the level related to concurrent atmospheric conditions.[17] The barometer seemed like a providential gift to the fellows' program to compile a history of the weather, a project advocated decades earlier by Francis Bacon. When Hooke's "Method for Making a History of the Weather" was published in 1667, it specified that barometer readings should be recorded alongside the measurements of thermometer, hygrometer, and wind gauge.

In these years, the Royal Society became a clearinghouse for reports of instrumental measurements and their relation to weather conditions. Oldenburg, the society's secretary, coordinated the efforts of a far-flung circle of correspondents and published some of their findings in the *Philosophical Transactions*. Beale, who lived in Yeovil in Somerset, was supplied with a barometer from London in 1664; his observations began to appear in print two years later. Henry Power, a Yorkshire physician and virtuoso who had been working on the Torricellian experiment since the early 1650s, was sent instruments to secure his participation in the project. The antiquarian and astrologer Joshua Childrey was invited to contribute his weather observations to the society and was sent a thermometer, though he reported that the first one dispatched from London had broken in the wagon bringing it to Salisbury. Oldenburg also enlisted the Continental savants Adrien Auzout (in Paris) and Johannes Hevelius (in Danzig), instructing them how to prepare their barometers and how to report their observations. The aim of the project, Oldenburg reported to Boyle in 1665, was to discover "whether ye Air gravitates more in ye East or West, North or South; towards ye Sea or Land; in hotter or colder weather, and keeps yesame seasons of changes."[18]

Notwithstanding all this activity, the connection between barometric variation and changes in the weather remained far from clear. In September 1663, Hooke somewhat prematurely declared that he had observed Boyle's barometer to rise in summer and fall in winter—a variation he ascribed to greater air density in the summer owing to increased exhalation of vapors from the earth.[19] This theory could perhaps account for the descending mercury level often observed to coincide with the discharge of vapors in the form of rain, but it encountered an obstacle when many observers reported that the mercury stood at a high level during winter frosts as well as during summer heat waves. Beale expressed his puzzlement that the barometer sometimes seemed to remain steady while the weather changed quite significantly. He wondered whether perhaps he had missed something through not watching it constantly.[20] At times, the device seemed particularly fickle and paradoxical. Who would have expected, for example, that the air would be shown to be heaviest when serene and lightest when laden with threatening clouds? Reviewing several years' research in 1666, Boyle concluded that it was "more difficult, than one would think, to settle any general rule about the rising and falling of the *Quick-silver*." He appealed for more observations by "divers Ingenious men," to illuminate "the odd *Phænomena* of the *Baroscope*, which have more hitherto pos'd, than instructed us."[21]

As inquiries continued over the following decades, little was accomplished to clarify the situation. A few correlations between the barometer's indications and weather patterns seemed to hold: the mercury stood high during calm, fair conditions and during winter frosts; a fall in the level generally presaged rain or a storm. But even these patterns were found to be not without exceptions. At the same time, debate continued as to the cause of the variation. In 1679, fellows of the Royal Society discussed whether a high level of the mercury column really signified an increase in the air's density or simply showed that the atmosphere was piled up to a greater height at that place. In 1684, the naturalist Martin Lister went so far as to doubt that the variation primarily reflected a change in air pressure. He speculated that it revealed instead the peculiar way mercury contracted and expanded as the temperature changed.[22] The following year, the *Philosophical Transactions* saw the publication of a theory that ascribed the variation to atmospheric effluvia. George Garden, a scholar and divine at King's College, Aberdeen, suggested that the weight of the air caused it to vary in its capacity to hold water vapor: heavier air had a greater capacity, and air that was losing weight tended to release its vapor in the form of rain. Variations in the density of the atmosphere in different places caused winds to flow between them. As to what caused the change in the air's weight itself, Garden suggested that it was due to its infusion with particles of some more subtle medium or "nitrous steams." John Wallis, the Savilian Professor of Geometry at Oxford and one of the leading mathematicians of the age, broadly endorsed Garden's theory. He agreed that variation in the pressure of the atmosphere was primarily due to the concentration of effluvia dissolved in it, along with the effects of heat and cold.[23]

In 1686, an alternative to the effluvial theory of the barometer was proposed by Edmond Halley, recently elected Clerk of the Royal Society, who had made his name by astronomical and meteorological observations during a year's sojourn on the island of St. Helena. Halley's paper in the *Philosophical Transactions* promised "an Attempt to discover the true Reason of the Rising and Falling of the Mercury, upon change of Weather." He began by listing eight generalizations concerning the barometer's behavior, which he claimed were "sufficiently known already to those that are curious in these Matters."[24] They included the mercury generally standing high during calm spells or frosts and falling during wind or rain, and the fact that the instrument varied much more in northern latitudes than in the tropics. He went on to try to explain these facts by invoking the influence of the winds surrounding the instrument. Converging winds would increase air density; diverging ones would lessen it. The loading of the air with vapors was almost entirely dependent on its density, Halley claimed. Less dense air released vapors in the form of rain; more dense air absorbed them into the atmosphere and held them there. Far from being *caused by* variation in density, the winds were independent *causes of* that variation, in Halley's account. The winds themselves were primarily caused by the physical forces set up by the earth's rotation and the heating effects of the sun. At least between the tropics, the winds tended to follow regular paths across the oceans, of which Halley, in a subsequent paper, provided a map.[25] It was perhaps the breadth of vision suggested by Halley's map of the trade winds that captured the attention of many readers. As W. R. Albury has noted, his theory was a "rather incoherent" one, but it earned Halley recognition as a leading authority on the barometer in the following decades.[26] By proposing a series of simple rules for understanding it, and gesturing at its potential utility for global navigation, he gained unique status as an interpreter of the instrument to his contemporaries.

The philosophers' dispute about the causes of barometric variation was echoed, though sometimes in a rather garbled way, by the writers of texts for the general public. The London clockmaker John Smith, in his *Compleat Discourse of the . . . Quick-Silver Weather-Glass* (1688), followed Halley in ascribing fundamental importance to the winds as the causes of variation in the air's density, although by the time he brought out his *Horological Disquisitions* in 1694, he had reverted to the view that changes in the temperature and moisture content of the air were probably more important. Richard Neve, author of another layperson's guide to the instrument, *Baroscopologia* (1708), declined to enter into what he called "philosophical niceties" but declared that it was primarily moisture content that determined the air's weight.[27] Writers closer to the leading experimental philosophers largely swung to Halley's side of the issue. John Harris, in his *Lexicon Technicum* (1704), quoted Halley's theory, as did John Theophilus Desaguliers in his *Lectures of Experimental Philosophy* in 1719. When Edward Saul, former Oxford don and rector of Harlaxton in Lincolnshire, produced his *Historical and Philosophical Account of the Barometer* in 1730, he dismissed Wallis's effluvial theory fairly summarily and declared for Halley's theory of the winds.[28]

By the beginning of the eighteenth century, the basic identity of the barometer as an instrument had been solidly established. Craftsmen involved in making the devices clearly distinguished them from thermometers or the traditional weather glasses. Smith wrote in 1694 that the latter were of very limited utility. Gustavus Parker, inventor of a new design of portable barometer in 1700, declared that the older artifacts were sources of "the greatest confusion that can be."[29] At the same time, however, there was not universal agreement that what the barometer showed was the pressure of the air. As late as 1674, the jurist Sir Matthew Hale was disputing the Royal Society's view that the Torricellian experiment should be understood in terms of air pressure.[30] Sir Samuel Morland, Master of Mechanicks to Charles II and inventor of an ingenious arrangement for magnifying the motion of the mercury column with a balance arm, pointer, and scale, apparently persuaded the king in 1677 that air pressure was not responsible for the movement.[31] Half a century later, the barometer maker and retailer John Patrick claimed to have shown Sir Isaac Newton that the behavior of his instruments could not be explained by the doctrine of air pressure.[32] It seems that even those most closely connected with making barometers were not uniformly convinced as to what their behavior showed.

When it came to the connection between barometric variation and the weather, disagreements were even more fundamental and persistent. A French commentator in the 1720s judged that all "Philosophers" agreed that the apparatus displayed changes in the weight of the air, but they did not agree what caused those changes.[33] In the late eighteenth century, writers of guides on how to use the instrument tended to strike the same notes as their predecessors a hundred years before. They acknowledged that most users wanted their barometers to predict the weather, while emphasizing that no certainty could be achieved in this matter. They suggested rules for interpreting the motion of the mercury in the tube, while stressing its fallibility as an indicator of what the weather was going to do. They called for more systematic research, suggesting that yet more data might allow for the derivation of laws of atmospheric phenomena. At the same time, the question of what caused the variation in the mercury level remained an open one among specialist experimenters. At a meeting of the Coffee House Philosophical Society, in London in December 1784, the issue provoked a debate that reenacted the basic arguments laid out a century earlier. The Swiss physician Aimé Argand defended the view that the concentration of vapors in the air was the most important determinant of its weight, while the writer and teacher William Nicholson ascribed primary importance to the winds. Either vapors or winds could be regarded as fundamental agents of changes in the weather. After listening to the discussion, the Irish virtuoso Richard Kirwan, currently serving as president of the society, declined to endorse either theory, while giving reasons to doubt both of them. The society's minutes recorded that everyone accepted that a true understanding of the causes of barometric variation remained a distant prospect, so that "none of the Gentlemen who spoke pretended to consider the subject otherwise than as in embryo." [34]

Delivering himself of his considered opinion on the topic, in a paper presented to the Royal Irish Academy in 1788, Kirwan sorted the available theories of barometric variation into three classes: those that ascribed causal agency to changes in temperature, to the confluence or dispersion of winds, or to the concentration of aerial vapors. The first possibility could be dismissed conclusively; one might say the barometer could not be recognized as an instrument at all unless it was understood to measure a variable independent of temperature. The second (Halley's theory) had been called into question by work by the Swiss meteorologist Jean André Deluc. Deluc and his countryman Horace Bénédict de Saussure had also cast doubt upon the causal agency of water vapor, since their hygrometric measurements had shown that vapor concentration was not directly correlated with atmospheric pressure. The way was therefore open for Kirwan to put forward his own theory, a complex account of movements of air in the upper atmosphere due to heating by the sun and the earth's rotation. He envisioned the atmosphere as a kind of tidal ocean, its height rising and sinking as currents of air flowed from place to place. Reflecting new discoveries, Kirwan inserted into his model the agency of electricity and the recently identified "inflammable air" (hydrogen). Inflammable air in the upper atmosphere was being ignited by electricity, he claimed, as witnessed in the auroras common in polar regions. This complex pattern of atmospheric motions was reflected in the oscillations of the mercury in the barometer tube. Like Halley before him, Kirwan was reasserting the significance of the barometer as a metonym of the macrocosm; its motions were to be seen as a key to large-scale changes in the wider world.[35]

Kirwan also indicated two factors that had exacerbated the difficulties of understanding the barometer during the century before he wrote. On the one hand, the apparatus had been extraordinarily widely used, becoming a token of the accomplishments of enlightened Europe. Kirwan wrote that it demonstrated the sterility of ancient philosophy and the necessity of experimental investigation. It had "thus excited the more civilized nations to those pursuits, which have since rendered them as superior both in arts and arms to the more ignorant, as men are to brutes." On the other hand, however, it was precisely this popular influence that had led to excessive claims about the predictive ability of the instrument: "Great, however, as has been the success attendant on the study and application of the observations made by the help of the barometer, the bulk of mankind have always expected still more from it: Nothing less indeed than a prognostic of the weather."[36]

Kirwan was pointing to the central paradox of the barometer as an instrument of enlightened science: its successful distribution as an emblem of enlightenment was precisely the condition of its misunderstanding by "the bulk of mankind." It had been seized upon by experimental philosophers as one of the great achievements of their enterprise, symbolic of the break with classical philosophy and of the program to establish natural knowledge on a consistently empirical foundation. It had been widely accepted as such by its users, who associated it with "polite science" and purchased it to signal their cultural refinement. The price of this success, however, was a widespread expectation that, in Kirwan's view, constituted a misapprehension: people looked to the barometer as a reliable way of forecasting the weather. They placed a degree of faith in it that resembled—or was liable to become— plain superstition. In the history that Kirwan was sketching, the barometer had turned out to be a most equivocal tool of enlightenment, or perhaps an indication of the ambivalences inherent in the movement itself.

The Instrument Trade and Consumers

How had this paradoxical situation come about? The barometer and other meteorological instruments owed their success to a confluence of interests among three groups: the experimental philosophers who first investigated

their properties, commercial instrument makers who perceived and exploited their value in the consumer market, and the purchasers themselves.
As with other objects circulated in such "trading zones," the instruments
had rather different meanings for each of the participating constituencies.[37]
The barometer, for example, was seen by experimental philosophers as
a tool for exploring the phenomenon of air pressure. It came to be used
in systematic weather recording, but experts tended to be cautious about
its reliability as a predictive device. Instrument makers saw the potential
for marketing a philosophical curiosity and also played up its prognostic
capabilities. For many consumers, no doubt, the claim that the barometer predicted the weather was a crucial incentive to buy one. While some
purchasers used it systematically to compile weather diaries, most did not.
To many who bought it, the apparatus was a conversation piece, an obscure reflection of bodily sensibilities, and an oracle that might foretell the
future.

Barometers made the transition from laboratory apparatus to household
object remarkably quickly. In the 1660s, they were said to be "confined to
the cabinets of the virtuosi," but by the 1710s they were described as common "in most houses of figure and distinction."[38] Neve, in 1708, declined
to give instructions for making a barometer because, he said, "any one
may be furnished with it in *London*, at a cheaper rate than he can make it
himself." He remarked that "few gentlemen [are] without one of them,"
although "few of them understand its right *Management* and use."[39] By
the 1720s, one observer claimed that they were "very common, and in every
Body's Hands."[40] The lawyer and virtuoso Roger North gave an account
of how the commercial sale of barometers had begun in his "Essay of the
Barometer," written sometime in the first two decades of the eighteenth
century but never published. North ascribed a crucial role to his brother,
Francis North, Lord Keeper to Charles II and himself something of a virtuoso. Francis North, his brother reported, "thinking this instrument was
too much confined among the curious, and that if the use of it was made
more general, to which a shop exposition would tend, some catalogues of
analogies might at length come forth," instructed his watchmaker, Henry
Jones of Inner Temple Lane, to make them and offer them for sale. Jones
soon lost interest, however, so North turned to Henry Wynne, a maker of
clocks and mathematical instruments in Chancery Lane, to continue the
trade. By the time of writing, Roger North asserted, the barometer had become so common in shops and gentlemen's parlors that there was no need
to describe it.[41]

North's account probably refers back to some time in the 1670s; it evokes the fascination with the barometer in the court of Charles II at that time. North recorded that a wheel barometer—in which the motion of the mercury was linked to the movement of a pointer on a circular scale—had been placed in the king's bedroom. This is echoed in a story Boyle told about King Charles's enthusiasm for the instrument. According to Boyle, the king claimed to have been able to predict a storm on the day of a royal outing by consulting his barometer.[42] While there is no reason to doubt its accuracy, North's account does not tell the full story of the beginnings of commercial barometer sales. He admitted that craftsmen working for the Royal Society, such as Hooke's assistants Richard Shortgrave and Harry Hunt, had been selling barometers to the public even before his brother prompted the clockmakers to make the apparatus more widely available. Hunt, who succeeded Shortgrave as Operator to the Royal Society in 1676, made wheel barometers and thermometers, and supplied at least one instrument to the diarist Samuel Pepys. Thomas Tompion, probably the most renowned clockmaker of his day, who also produced instruments for Hooke and Boyle, published one of the earliest known advertisements for a barometer in 1677. By 1694, the clockmaker Smith was recording that many kinds of barometers were being sold in the London shops. Other members of the clockmakers' guild, including Thomas Tuttell and Daniel Quare, took up the barometer trade. Tuttell sold barometers alongside surveying and navigational equipment in his shop at Charing Cross. Quare became one of the leading makers and suppliers of the instrument around the turn of the eighteenth century, producing distinctive and innovative designs. The barometer also found a place in the shops of optical instrument makers, such as John Yarwell and John Marshall, rivals and neighbors of one another in Ludgate Street near St. Paul's Cathedral.[43]

Yarwell and Marshall may have obtained the barometers they sold from John Patrick, who emerged around 1700 as one of the first craftsmen to specialize in making this instrument. Patrick was originally apprenticed as a cabinetmaker; his good connections with the scientific community were matched by excellent craftsmanship and advertising skills. In the first decade of the century, he published the pamphlet *A New Improvement of the Quicksilver Barometer*, making him one of a number of craftsmen who claimed at the time to have redesigned the apparatus to make it more portable. He had earlier teamed up with the Clockmakers' Company to oppose the granting of a patent to Quare for his own design of a portable barometer. As Patrick built his business in premises in the Old Bailey, he was boosted by an endorsement from the coffeehouse mathematical lecturer John Harris, in the *Lexicon Technicum* (1704). Harris wrote that he had "never seen

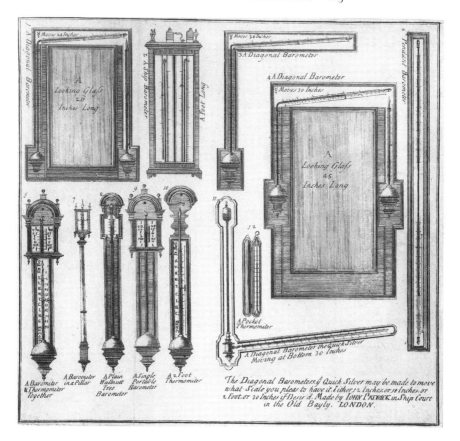

FIGURE 12 · John Patrick's trade card, c. 1710. This advertisement by the first specialist barometer maker shows the designs of instruments he had available, including diagonal barometers mounted over mirrors. Banks Collection, Department of Prints and Drawings, British Museum. © Copyright the Trustees of the British Museum. Reproduced by permission.

better Weather-Glasses of all kinds made any where than by Mr. Patrick"; he recognized in the virtuoso craftsman a kindred experimenter and fellow entrepreneur.[44] Patrick was circulating his own advertising trade card, offering his customers an upright-tube, open-cistern barometer for two guineas. Most of the cost probably went into the wooden case: oak veneered with walnut was typical, with elaborate carved scrolls as ornamentation or even statuettes of Greek gods mounted on the top. Patrick also advertised an apparatus with a diagonal tube at fifteen guineas, in which the barometer was bent into an inverted L, mounted around a mirror. The motion of the mercury was magnified along the length of the top segment of the tube,

FIGURE 13 · Diagonal barometer with thermometer and calendar. Instrument made by Watkins and Smith, London, c. 1763. The diagonal barometer is combined with a thermometer and hygrometer, and in the center is a perpetual calendar. Courtesy of Science and Society Picture Library, London.

which was inclined at an angle just above the horizontal. Patrick explained that the mirror was included so that "Gentlemen and Ladies at the same time they Dress, may accommodate their Habit to the Weather.—An Invention not only Curious, but also Profitable and Pleasant." [45]

Other versions of the diagonal apparatus came without a centerpiece, or with an almanac or calendar in the middle, but the conjunction of the barometer with a mirror intrigued many customers. Apart from the convenience of allowing people to adjust their dress for the expected weather, the mirror and the barometer both offered public reflections of private identities. As Terry Castle has noted, the two devices provided different sorts of reflections of the self.[46] The barometer registered changes that were also experienced by the individual as alterations of health or mood; it supplied an external representation of internal feelings. As they contemplated their image in the mirror, people could also calibrate their own emotional state by reference to an instrument that displayed the changing qualities of the air. Both mirror and barometer were therefore appropriately sited at the point where the individual composed herself or himself before emerging into the shared space of the public sphere. Even before this kind of combination apparatus became popular, George Sinclair was reporting in 1688 that "Ladies and Gentlewomen at *London* do Apparel themselves in the Morning by the Weather-Glass." [47]

Patrick's example as a specialist barometer maker and retailer was followed by some other tradesmen. More commonly, the devices became part of the regular inventory of clockmakers, opticians, and makers of mathematical instruments. London tradesmen who included barometers on their advertising cards included Steven Davenport, who sold mathematical and philosophical instruments in Holborn in the 1720s and 1730s; Edward Scarlett, who made optical instruments in Soho from around 1700 to 1740; and Thomas Blunt, operating in Cornhill for about five decades from the 1770s. These and many other London merchants supplied a growing provincial demand; their products were retailed by apothecaries, spectacle makers, and clockmakers throughout the country. They were also joined by dozens of provincial barometer makers in the course of the eighteenth century, including opticians, jewelers, and gunsmiths, along with many clockmakers. Sinclair was selling barometers in Glasgow while also serving as a professor of mathematics at the university in the 1690s. In the following century, a series of Scottish makers established themselves in Glasgow and Edinburgh. John Hallifax started out as a clock and barometer maker in Barnsley, Yorkshire, in 1711; he was succeeded by his son George, who continued the business in nearby Doncaster for the remainder of the century. Even quite small towns sometimes had native barometer makers. The brothers John and William Bastard were selling wheel barometers in Blandford, Dorset, from the 1720s. Charles Orme opened his business selling diagonal barometers in Ashby-de-la-Zouch, Leicestershire, around the same time.[48]

Patrick also showed many tradesmen the way to promote barometers by bolstering luxury design elements with scientific credentials. Expensive wood veneers continued to be used, imported mahogany replacing walnut around the 1730s. The architectural flourishes favored by Patrick were adopted by many other makers; experts can still date surviving specimens and identify their makers by the changing fashions in pediments and scrollwork.

Patrick's pricing seems to have been fairly typical for top-end models: customers were happy to pay several pounds for a luxury item that would assume a prominent place in their home. Barometers, in this respect, were comparable to expensive pieces of furniture made by the best designers; they were fashionable consumer goods in an age of conspicuous consumption, when domestic furnishings were chosen to display the affluence and social standing of the householder.[49] But, of course, barometers were also sold as scientific or "philosophical" artifacts. Patrick made sure that customers knew that his novel pendent barometer had been "Examin'd and Approv'd by several Persons of Quality of the ROYAL SOCIETY." [50] The apparatus was regularly described by its promoters as a philosophical "wonder" or "curiosity," and its credentials were warranted by mention of the "many Ingenious and exquisite Searchers into Nature" who had studied it.[51] In 1766, the entrepreneurial lecturer and instrument maker Benjamin Martin described the barometer as "both in Regard of Curiosity and Utility, . . . the first in Dignity among the modern Philosophical Inventions." [52] The names of Torricelli, Boyle, and Hooke were routinely invoked in the promotional literature, and even "the GREAT Sir *Isaac Newton*" could be summoned to give his blessing, since, as John Laurence explained in 1718, the movement of the mercury had to be explained by reference to the theory of gravitation.[53]

This dual identity—as philosophical instrument and desirable luxury—was the key to the barometer's success. The instrument was used with care by serious weather observers, from William Derham at the end of the seventeenth century to Luke Howard at the beginning of the nineteenth. In the 1750s, William Borlase, living near Penzance in Cornwall, meticulously compared the readings of his diagonal barometer with an upright instrument. He concluded that the former was less quick to respond to changes in atmospheric pressure because the mercury was less free in its movement.[54] Borlase, like Gilbert White a couple of decades later, understood that barometers needed to be corrected for the height of observation above sea level. White noted the discrepancy between his instrument and that possessed by a gentleman neighbor whose house was three hundred feet higher in elevation.[55] A similar concern with exactitude was shown by White's friend Thomas Barker, who used a barometer designed by Quare, with a sliding vernier scale that allowed measurements of the height of the mercury column to one hundredth of an inch.[56] For most users, however, such precision was irrelevant. The barometer had a fascination greater than that of any mere measuring instrument, because of its mysterious responses to imperceptible changes and its apparent ability to foretell the future. Howard understood that this was the key to its popularity. He wrote in 1820 that "the elegance of its construction, the facility of observing its changes, perhaps also something mysterious and imperfectly understood in its indications, have made this instrument but too successful a rival to the Thermometer." [57] The thermometer might have had more potential as an instrument of quantitative meteorology, but for most of the eighteenth century, it was the barometer that captured much greater public interest.

The barometer's unique appeal for consumers was as a luxury household object that was also a philosophical curiosity. Those who bought it were buying into the experimental sensibility and acquiring a symbol of intellectual refinement. The device demanded at least some degree of manipulation from its owners. Purchasers of the early versions were expected to fill the tube with mercury themselves, taking precautions to clean the glass and avoid getting air bubbles in the column. [58] In 1747, the *Universal Magazine* reiterated the standard advice on this procedure and went on to give instructions on replenishing the mercury regularly. John Warner, an instrument maker active in London around the turn of the seventeenth and eighteenth centuries, published a blank grid on which barometer readings and other weather observations could be recorded. [59] Thus, purchasers were invited—though obviously not required—to make regular records of the weather. The instrument gave its owners a symbolic link with enlightened science, even if it did not require them to make measurements to a high standard of precision. It was appropriately labeled a "Philosophical or Ornamental Branch of Furniture," [60] because its status as an ornament of the bourgeois home was integrally bound up with its philosophical credentials.

Interpreting the "Oraculous Glasses"

Edward Saul, who coined the phrase "Philosophical . . . Furniture" to describe the barometer, remarked on its role in domestic conversation, on the way it often supplied "Matter of Discourse upon the various and sudden Changes of it." [61] As Samuel Johnson pointed out, British people talked about the weather so much because it changed so often. The barometer became a conversation piece because it seemed to offer a key to these changes; it gave mundane discourse about the weather a new inflection, bringing an "air of philosophy" into the parlor or the drawing room. In Marcellus Laroon's painting *A Musical Conversation* (c. 1760), the instrument was shown in an appropriately prominent location on the wall of a well-appointed drawing room, which was populated by a group of affluent men

FIGURE 14 · Henry Bunbury, "Club Night" (1785). The diagonal
barometer on the wall, with its indication of "stormy," seems
to reflect the edgy relations between the characters. Courtesy of
The Lewis Walpole Library, Yale University.

and women enjoying music and talk.[62] In Henry Bunbury's cartoon "Club
Night," it accompanied a group of men informally smoking and chatting.
The barometer belonged at least as much in such company as in a gentle-
man's library or a philosopher's laboratory.

Benjamin Martin dramatized the kind of domestic conversation that
might occur around the barometer in his *Young Gentleman and Lady's Philosophy* (1772). A young man, Cleonicus, returns home from university to
converse with his sister, Euphrosyne, whose education has been confined to
the home. The barometer, he tells her, shows changes in the air that affect
individuals' health and fortunes; merchants, mariners, and husbandmen
know this well. In the world beyond the sheltered domestic sphere, Cleonicus tells his sister, the import of the apparatus is "a Matter of more than mere
Pleasantry." Knowledge of imminent weather conditions can be of significant economic value, and the barometer has the ability to "prognosticate the
future Changes that may immediately happen.... On such Fore-knowledge
you will readily allow, a great deal must depend." Euphrosyne concedes
the point, expressing astonishment that not everyone possesses such an in-

dispensable instrument. "Who would, for the Sake of such a small Sum of Money, want a general Index for Life, Health, and good Fortune?" she asks.[63]

Martin's little scene suggests that the barometer could be used to reinforce male authority in the middle-class household. The instrument, said by the author to be particularly suitable for "a Gentleman's Parlour or Study," provided opportunities for men to display their sophistication to women, who would be expected to have less in the way of scientific education and experience of the world. Saul, too, envisioned that owners of the apparatus might be "desirous of exerting now and then a superiority of understanding, by talking clearly and intelligibly upon it."[64] But the barometer was an equivocal asset for buttressing masculine authority, since its mercurial unreliability was frequently described in feminine terms. Martin was obliged to admit that the device was tainted with female folly insofar as it was an object of fashionable consumer desire. He explained that many purchasers were seduced by expensive and needlessly elaborate models, such as the diagonal-tube versions that gave an illusion of greater accuracy. Consumers were led astray because "many Gentlemen, as well as Ladies, affect to have Things very fine and showy, and such as shall make a grand Appearance."[65]

The aura of feminine folly surrounding the barometer was directly connected with its use as a means of prediction. Roger North remarked that the apparatus was already beginning to be "lookt upon as a piece of female incertainty" by the time he was writing.[66] Many observers concluded that to place too much faith in its forecasts was to surrender one's reason to passion or credulity. Johnson was particularly scornful about this. In his periodical *The Idler*, on 5 August 1758, he wrote:

> The rainy weather which has continued the last month, is said to have given great disturbance to the inspectors of barometers. The oraculous glasses have deceived their votaries; shower has succeeded shower, though they promised sunshine and dry skies; and by fatal confidence in these fallacious promises, many coats have lost their gloss, and many curls been moistened to flaccidity.[67]

Describing the objects as "oraculous glasses" with their own "votaries," Johnson was clearly invoking the aura of superstition and primitive prophecy. He professed to regard all such fallacious schemes of divination as trespasses on the domain of divine providence. The mention of glossy coats and curls associated this gullibility with feminine attributes; but clearly, this was not to say that only women were guilty. Men — even educated men — might so far abandon their masculine powers of reason as to succumb to the same folly. The lesson was reinforced a few months later, in *The Idler* of 2 December 1758, when Johnson presented the "Journal of a Senior Fellow." This purported to be a series of extracts from the diary of a Cambridge don, whose unproductive life and hypochondriac sensibility were regulated by a neurotic degree of attention to his barometer.[68]

Others also recognized the problem—even North, who admitted to being a "votary" of the instrument himself. Early in the century, North wrote that the barometer had already suffered a loss of status because the "vulgar" had invested excessive hopes in its predictions. It was scarcely surprising that most people looked to it for prophecy; the apparatus had become public precisely because it was said to have prognostic virtues. And the human propensity to try to know the future was surely universal. As North asked, "of all human kind who is he that lives, and is not extreamly concerned to obtaine (if possible) A certain presage of weather[?]"[69] It was to gratify this demand that the very first makers of commercial barometers had begun the practice of engraving weather indicators on the scale beside the mercury tube. On British instruments they quickly assumed a familiar order, reading from the top "very dry," "settled fair," "fair," "changeable," "rain," "much rain," and "stormy." The emphasis on rain seems symptomatic of British concerns; Continental scale markings put rather less emphasis on this feature. On some models, separate scales for summer and winter were mounted on either side of the mercury tube, reminding people that high pressure usually accompanied settled heat in summer and hard frost in winter. Experts regularly stressed that the scale markings were far from infallible (North described them as "like a cheating oracle that . . . pretends to tell fortunes"),[70] but they admitted that most users trusted them. Instrument owners were often advised that the motions of the mercury were more significant than where it stood against the scale. Neve told his readers, "Most *Gentlemen* that have *Weather-Glasses*, not knowing . . . this Rule; always expect such *Weather* as the Words (against which the mercury stands) express, and thereby deceive themselves and others."[71] But such deception was unsurprising, when the prognostic signs could apparently be read directly from the words on the instrument itself. The scale markings sent a clear message to most users of the barometer that the weather could be straightforwardly predicted by reference to them.

The status of the barometer as a kind of oracle was consolidated early in its history by an association with astrology. Astrological almanacs welcomed the public appearance of the device in the late seventeenth century, promoting it in their pages as a tool for weather prediction.[72] In 1700 Gustavus Parker, apparently an associate of the instrument maker John Yarwell, set out to promote the prognostic abilities of a new design of portable barometer. His promotional pamphlet made grandiose claims on behalf of the new instrument but gave no details of its design. Parker asserted that his new invention had enabled him to discover the true cause of the tides, of magnetic needles' pointing north, and of winds, rain, snow, and other meteorological phenomena.[73] He also published a forecast of the weather for the month of September 1700, which he claimed to have calculated from the indications of his instrument, and followed this up by issuing further monthly or fortnightly predictions during the following year.[74] Provoked by Parker's remarks against traditional astrology, John Gadbury responded with a tract insisting that stars and planets were still "the best barometers." Gadbury resented the suggestion that the barometer could displace planetary weather forecasting, particularly because—as he pointed out—Parker was himself an astrologer.[75] For Gadbury, the instrument was a complement to celestial signs of coming changes in the atmosphere; it could give valuable indications of the weather for a day or two at a time, but it could not rival the planets as predictors for a coming season: "Such Philosophical foresight belongs not to the *Barometer-Men*, as such, but to the *Astrologers*."[76]

Parker's predictions were also taken to task by John Patrick, at this point just beginning his career as a barometer maker. Patrick reissued Parker's prediction for September 1700, juxtaposing the forecasts for each day with his own observations of the weather that had actually transpired. On twenty-five out of thirty days, according to Patrick, the prediction had missed the mark. Parker had grossly overstated the prognostic ability of his instrument. No barometer, Patrick insisted, "will fore-shew the Weathers Alterations a Month or Six Weeks, as he does Allege."[77] Recalling the episode a few years later, Roger North dubbed Parker's monthly predictions a "cheat," which happened by chance to be moderately accurate for the first month but quickly failed thereafter. The whole event was testimony to the "generall disposition of folks to crye up wonders," and Parker's pamphlet about his invention was "such a jargon of nonsense, as one would have thought it calculated onely for Idiots to read."[78] North's comments echo his other remarks about the difficulty of disentangling the barometer from superstitious expectations in the public mind. The device clearly was a "wonder" of some kind, and it was hard to persuade people that it was not comparable to astrology as a source of seasonal predictions. Patrick cautiously declined to condemn astrology as such, declaring that he left the matter to the astrologers themselves, while noting that some of the almanacs had proven more accurate in their predictions than Parker.

Authors of astrological almanacs continued to describe the barometer

FIGURE 15 · James Gillray, "Very Slippy Weather" (1808).
Another from Gillray's series on weather and character.
The implication seems to be that the barometer does not forewarn
of all the weather hazards one might encounter. Courtesy of
The Lewis Walpole Library, Yale University.

sympathetically throughout the eighteenth century. Francis Moore's *Vox Stellarum* for 1791, for example, suggested that the instrument could be used "in Conjunction with the Weather given in this Almanack by Philosophical and Astrological Rules" to predict conditions for the season ahead. It was understood that neither astrology nor the barometer gave certainty, but both yielded indicative signs that could be compared with one another to increase the likelihood of a correct result. The publication provided a series of fourteen rules, tying changes in the mercury level to the surrounding circumstances of wind, temperature, weather, and season. By weighing all these factors together, it was proposed, the interpreter could learn to read the barometer's prognostic signs.[79]

Whether they connected it with astrology or not, most users seem to have interpreted the barometer by reference to rules of this kind. Many writers offered maxims that specified how the instrument could be used for weather prediction. In this sense, interpretation of the barometer as a kind of oracle was pervasive among those who commented on the instrument. It was widely regarded as yielding prognostic signs that required interpretation by reference to other signs, including many of those inherited from traditional weather lore. Halley gave eight such rules, some of which required the observer to take account of wind direction, season, or temperature to interpret the device's behavior. Patrick published ten rules of a similar kind, and Neve fourteen. Harris's *Lexicon Technicum* indiscriminately listed Halley's rules and Patrick's, while Philip Miller's *Gardener's Dictionary* (1752) continued the trend by adding eight rules by John Pointer and sixteen by "another author." [80] Directions to attend to the way the mercury was moving and where the wind was blowing were very common, and rules were formulated to work out what weather should be expected given the conditions at different times of year. Peter Rabalio, an instrument maker operating in Birmingham in the late eighteenth century, published a handy sheet summarizing his rules, which he derived from those of Halley and Patrick. His final remark suggested that the barometer could be used to forecast wet or dry seasons at the times of the equinoxes, an indication that the instrument was still widely looked to for seasonal predictions.[81]

Some of these maxims referred to the kind of signs traditionally recognized as part of the hermeneutic practices of weather-wising. As John Campbell wrote in his edition of John Claridge's *Shepherd of Banbury's Rules* (1744), "The Art of prognosticating the Weather may be considered as a Kind of decyphering." [82] Forecasters were accustomed to looking around them for the indications of changes in the weather; they decoded them with reference to proverbs and other oral lore handed down by tradition. Campbell himself was guarded about the utility of the barometer in this kind of practice. It was a "wonderful invention," he wrote, "curious" and "ingenious," but it had yet to prove itself in long-range forecasting.[83] Other authors more readily assimilated it to traditional predictive methods. John Smith, for example, recommended that the motions of the mercury should be watched along with the appearance of the sun, cloud formations, the phases of the moon, the wind's strength and direction, and a clutch of indicators from animal behavior: hooting owls, flying bats and gnats, cawing crows, and pigs "crying in an unusual manner." [84] The Reverend John

Laurence, in his *Fruit-Garden Kalendar* (1718), offered his readers a series of rules in which fluctuations of the mercury column were related to other, more traditionally recognized signs of the weather.[85] The same approach was followed by John Mills in his *Essay on the Weather* (1770). Mills was cautious in his appraisal of the barometer, but welcomed it as a supplement to more traditional means of weather prediction. "When the character of the season is once ascertained," he wrote, "the return of rain, or fair weather, may be judged of with some degree of certainty in some years, though but scarcely to be guessed at in others, by means of the barometer." [86]

Like Mills, many commentators on the barometer judged that the apparatus could be used successfully provided that one appreciated its inherent uncertainties. Laurence wrote that "it is undoubtedly the best *Guide* we have; and a *Guide* ought, and will not fail to be treated with Respect, so long as *Modesty* is preserved; so long as *absolute* Power and *uncontroulable* Dominion are not pretended to." [87] As Martin and others pointed out, while the apparatus could yield only probable predictions of the future, most human decisions had to be made on the basis of knowledge that was only probable, though none the less valuable for that.[88] The barometer's indications and the signs traditionally recognized in weather forecasting were to be seen as predicting the future with a comparable degree of uncertainty. Anyone who expected them to produce certain knowledge was surrendering to superstition. To scorn them, however, would be to turn one's back on the bounty of nature, which was providentially constituted to provide discernible guidance for human life. As the barometer maker George Adams, a leading member of the London instrument trade, remarked in 1790:

> There are many, indeed, among the pretenders to sagacity, who treat all prognostic signs of the weather with contempt; as fit only for the attention of rude and uncultivated minds. They may, however, be told, that the processes continually carrying on by nature, on every side, and which are obvious to every eye, are as much the instruments of knowledge, as the more refined apparatus of the experimental philosopher.[89]

Adams's comments pointed to the circumstances in which the barometer had been successful in the century prior to his writing, and to a source of the difficulties surrounding it. The apparatus had been accepted as a gauge of atmospheric properties that could not be directly apprehended by the senses. It was understood to give indications of approaching changes in the weather, indications like the natural signs long used in forecasting. Adams himself situated the instrument in a tradition of weather signs going back to the ancients. With their catalogues of significant phenomena, Aristotle, Pliny, and Virgil had "given proofs of genius, that have not been equalled by all the efforts of succeeding ages." They had shown the way to increase the reliability of predictions by widening the range of signs taken into account. A single apparatus could not forecast reliably, so all available instruments should be consulted, together with indicators from nature. Adams acknowledged that most of those who purchased instruments intended to use them "not so much to know the actual state of the elements, as to foresee the changes thereof." In order to improve the instrument's predictions, it was necessary "to multiply observations on as great a number of signs as possible." [90] It was within this kind of hermeneutic practice that the barometer had achieved recognition as an oracle of the weather.

Assimilation of the barometer in the context of this sort of forecasting raised problems, however, because of course the apparatus was not a "natural" entity. Many commentators were inclined to diminish its significance for this reason. If the best weather signs came from nature, then what need was there for a human artifact to supplement them? Campbell and Mills both proposed that animals, birds, and insects instinctively apprehended changes in the air that were only indirectly and crudely represented by the barometer. And a number of medical writers suggested that since the human body was itself a sort of natural barometer, it would be a more sensitive indicator of atmospheric changes than the artificial instrument. The Scarborough physician Thomas Short proposed in 1750 that barometers might be useful for city dwellers, who lacked the countryman's proximity to nature, but the former would be entirely wrong to despise the lore of the latter "as a Cloud of foolish popular Predictions from the Brute and Vegetable World." [91] As in Adams's words, already quoted, those who ridiculed the "rude and uncultivated minds" of the rural population might themselves be justly criticized for their ignorance. Decades earlier, Roger North had already asked whether the new experimental science might allow philosophers to become as good at weather prediction as uneducated sailors and farmers. He professed, with gentle irony, to see "no reason, but a gentleman might in this Respect be as wise as his husbandman, or shepheard, if he would but attend ye phenomena of ye air." [92]

Because of these ironies and ambiguities, the barometer was a quite equivocal symbol of enlightened science. While it owed its origin to the labors of experimental philosophers and retained something of their intellectual prestige, it was received by middle-class purchasers as a means of predicting the weather. This required people to integrate the apparatus into traditional practices of weather-wising by comparing its prognostic signs with proverbial lore. But to evoke traditional practices was to suggest that nature would always be a more authentic source of weather knowledge than any artificial contrivance. The popular success of the instrument was based on its association with long-established means of weather prediction, but the connection often worked to undermine its credentials. The barometer, it seemed, could never be as good a predictor as the natural weather signs that were used to interpret it. For some commentators, at least, it was a reminder of—and scarcely adequate compensation for—humanity's departure from a natural mode of life.

Furthermore, the continuing mysteriousness of the device's behavior elicited attitudes in the population at large that middle-class intellectuals viewed as superstitious. An anonymous commentator in the 1790s, reviewing the history of the apparatus over the previous century, recalled the extravagant attempts to draw predictions from it for a month at a time, "than which nothing more absurd could have been imagined by a people in the rudest state of barbarism and ignorance." The lapse into primitive credulity had provoked others to denounce the instrument as completely useless for forecasting—an equivalent irrationalism from the opposite side. This writer concluded that "truth in this case, as it generally is, will be found between the extremes."[93] But the right balance was, as always, difficult to strike. The barometer seemed symptomatic of an enlightened reason threatened from all sides by passions, faith, and primitive superstition. It signified by its mercurial oscillations the turmoil of human emotions under the influence of atmospheric forces, and it was itself the site of passionate investment as an object of consumer desires. An instrument of polite science, it was always liable to be mistaken for a kind of magic. The tendency to place excessive faith in the predictions of the instrument was regularly castigated, but also tacitly encouraged by the way it was marketed. The barometer came to represent the limits of human knowledge as much as its triumphs, the less than universal reach of enlightened reason, and the troubling survival of beliefs and attitudes supposedly consigned to the past.

Sensibility and Climatic Pathology

The ill effects of bad weather appear often no otherwise, than in a melancholy and dejection of spirits, though without doubt, in this case, the bodily organs suffer first, and the mind through these organs.

EDMUND BURKE · *A Philosophical Enquiry into the Origin of Our Ideas of the Sublime and the Beautiful*

I have been lying on the sofa in a state of utter torpor. I mean to go out today to see if I am well or not. . . . If the present beautiful weather continues I shall be compelled to go and be happy in the country but at present I prefer being miserable in London.

ERASMUS ALVEY DARWIN · letter to *Frances Wedgwood*

MANY PEOPLE IN EIGHTEENTH-CENTURY Britain were concerned about the influence of the atmosphere on human health. A prominent aspect of the age's deepening interest in the weather was the question of how it affected bodily and mental well-being. The British believed that their climate had its characteristic virtues, but they also knew that the air sometimes made people sick. This susceptibility had been studied since ancient times, going back to the Roman physician Galen, and before him to Hippocrates, the supposed father of ancient Greek medicine. By the early eighteenth century, some of those compiling diaries of the weather were also beginning to record the prevalence of diseases in their localities. The new meteorological instruments of the period were used to investigate the physiological effects of temperature, moisture, and atmospheric pressure. In the late eighteenth century, a new instrument—the "eudiometer"—was introduced to measure how breathable the air was. These inventions took the place of the seventeenth-century weather glasses, providing instrumental indicators of the human body's mysterious sensitivity to the qualities of the air.

There had long been individuals who saw themselves — or were seen by others — as especially vulnerable to the atmospheric environment. The Worcestershire diarist of 1703 is a good example of a self-identified "melancholic," whose valetudinarian anxieties made him particularly subject to the weather. In the late seventeenth century, such a person risked being labeled an "enthusiast." In the early decades of the eighteenth century, climatic susceptibility began to be seen as a more widespread social problem. People worried that more individuals were succumbing to aerial pathologies in the conditions of modern life. Melancholia was sometimes thought to be especially prominent among the British. Notwithstanding the national pride in the qualities of the air, it was acknowledged that the prevailing dullness and dampness could have a depressing effect on the spirits of the population. Medical writers also pointed to modern luxuries as causes of increased atmospheric susceptibility. Fashionable clothing, indoor entertainment, and the consumption of tea and coffee were all thought to be making people more vulnerable. It was believed that sensitivity to aerial maladies was increased by the debilitating effects of luxurious living. Apparently, the diseases of the air were also, to some extent, diseases of modern life.

In this way, climatic susceptibility came to be seen as an index of social and cultural change, another "barometer of Enlightenment." It was viewed as the unfortunate consequence of certain trends in British society, especially the cultivation of personal sensitivity in manners, moral behavior, and aesthetics. As a number of historians have noted, this period saw the rise of the "culture of sensibility."[1] In polite circles, a heightened sensitivity to the feelings of others and to the beauty of one's surroundings was validated and encouraged. The civilized individual was expected to have refined manners and fastidious tastes, and to be motivated by an empathy for his or her fellow creatures. Sensitive feelings were prized as the basis of morality and aesthetics, but they were also thought to make people vulnerable to the disturbing effects of their environment on health or mood. It might be admirable to feel deeply in response to literature and music, to respond emotionally to the sufferings of other people and animals, but it was all too easy for refined feelings to become a kind of pathology. People were thought to be getting sick because they had become too sensitive to the influence of things around them, including the air.

The issue of sensibility and its effects was both a moral and a political one. Some moralists regarded the indulgence of personal feelings as a lapse of self-control, often characterized in gendered terms as a surrender of the masculine powers of reason to the feminine passions. For an individual, intemperance or self-indulgence could leave one vulnerable to diseases, including those originating in the air. Loss of rational control was also a problem for society at large, leading in the extreme to manifestations of mass enthusiasm, collective insanity, or religious frenzy. Writers such as Samuel Johnson and Thomas Short saw atmospheric susceptibility as a general loss of mental autonomy, portraying it as a sign of social corruption under the influence of luxury and soft living. For these conservative-minded commentators, climatic sensibility was indicative of the contemporary moral decline into laxity and effeminacy. It was a symptom of the decadence of modern society.

Others saw atmospheric susceptibility as a motivation for progressive social change. They went beyond merely studying the weather to trying to control it—at least as regards the quality of the air within and around human habitations. This air was to be improved to make it more conducive to health. Architectural schemes and projects for urban renewal strove to increase ventilation and the quality of the air people breathed. In the 1770s and 1780s, new techniques of "eudiometry" were used to assess the quality of the atmosphere. Discoveries in the chemistry of gases were exploited to yield new methods of treating the sick, sometimes yoked to quite radical schemes for enhancing the quality of the air at large. Those advocating reform often shared with conservatives the notion that civilization had brought with it a specifically modern vulnerability to atmospheric diseases by removing people from a natural mode of life. Even writers who were sympathetic to enlightened progress worried about these effects. But in seeking to alleviate them by projects of scientific therapeutics and comprehensive social intervention, the reformers parted company with the conservative moralists. They saw bad air as a challenge that demanded more systematic measures of enlightened reform.

Toward the end of the eighteenth century, atmospheric reform was being pursued in the spheres of individual therapy and social development. The leading reformers were advocating new modes of medical treatment, along with improvements in the environment and in the institutions of medical practice. Proposals to change the air became the centerpieces of wide-ranging programs of social reform. In the fraught political climate of the end of the century, these programs became very controversial. They were portrayed as expressions of rationalistic hubris and even satirized as a new kind of atmospheric pathology. But social programs of environmental improvement survived the century as a crucial component of the inheritance of the Enlightenment. The eighteenth-century reformers of the atmosphere bequeathed to their successors a legacy of concern for the health effects of the air and a determination to address the problem by comprehensive social change.

The Hippocratic Revival

In modern times, Hippocrates has been described as a "name without a work," because the various texts ascribed to him since classical antiquity cannot be authentically connected with any historical individual.[2] In the eighteenth century, however, it was generally accepted that Hippocrates himself was the author of these writings, and stories about his medical accomplishments—for example, that he had halted the plague in Athens by having fires lit throughout the city—were widely reproduced. Central to the Hippocratic legacy was the idea that human diseases were caused by the physical environment and recurred regularly with the changing seasons. Works of the corpus, especially the *Aphorisms*, the first and third books of the *Epidemics*, and *Airs, Waters, Places*, traced the onset of disease to climatic conditions in the places affected. Physicians inspired by the Hippocratic tradition were supposed to be able to give advice on where were the healthiest places to live, paying attention to factors like the prevailing winds, the type of soil, the quality of the water supply, and the proximity of marshes and other sources of "bad air." They believed that if they knew enough about the normal conditions at a particular place and season, they could predict the diseases in the coming year. Illnesses—at least insofar as they were common in a population at a particular time and place—were viewed as products of the environment. Thus, the key to defeating them was the doctor's knowledge of the local climate and the pattern of its seasonal variations.[3]

By the eighteenth century, these core doctrines had been combined with other medical theories of ancient, medieval, and Renaissance origin. Galenic medicine recommended that people adjust their diet and regimen to their physical surroundings. Good health would result from achieving the right balance of what medieval writers dubbed the "non-naturals"—the circumstances that affected individual constitution, including air quality, exercise, sleep, nutrition, evacuation, and the passions. In the Renaissance, external macrocosmic influences were thought to bear upon the bodily microcosm.[4] The great sixteenth-century French jurist Jean Bodin grappled with the question of how the mind was affected by physical forces impinging on the body. He concluded that the soul itself, though "free from all materiality, yet . . . is very much influenced by the closeness of the association" with its corporeal dwelling place.[5] Robert Burton's *Anatomy of Melancholy* (1621) agreed that the air's action on the mind occurred by the mediation of the body. Mental disorders like melancholy often went together with physical ones, according to Burton, both being brought on by the effects of bad air.[6] Edmund Burke's assertion in 1757 that depression and melancholy are caused by the atmosphere's influence on the bodily organs reflected the lingering hold of this classical tradition.[7] Emotional responses to the weather were taken as testimony to the sway of the bodily passions over the mind, reminders of the human intellect's lifelong imprisonment in the material body.

Because diseases were regarded as seasonal entities in the Hippocratic tradition, medical practitioners were supposed to attend to the astronomical markers of the calendar. The "dog days," the hottest of the year, associated with the rising of the star Sirius just before dawn, were supposed to be critical for many medical conditions. Although the Hippocratic texts did not countenance the idea that stars and planets directly affected diseases, many Renaissance physicians firmly believed in such celestial influences. Astrological interests led many doctors to try to match the daily progress of their patients' illnesses to the movements of the heavenly bodies. The weather was an obvious mediator of these effects for those who saw celestial influences at work in the atmosphere. In his letter to Samuel Hartlib, published posthumously in 1692, Robert Boyle proposed that the heavenly bodies caused sickness by affecting the properties of the air and its contamination by earthy effluvia. To study these effects, Boyle recommended the compilation of journals of the weather that would also record the motions of the heavens, "it being much more commendable for a Man to preserve the History of his own Time, . . . than to say, upon every Occasion that offers it self, this is the hottest, or this is the coldest; or this is the rainiest, or this is the most seasonable or unseasonable Weather that ever he felt; whereas it may perhaps be nothing so."[8]

Systematic recording of the weather was recommended to correct this sort of casual vagueness. Boyle believed that the daily journal would yield more reliable information about the atmosphere and lay bare its physiological effects. The same hope was shared by other members of the early Royal Society. John Locke appears to have undertaken his weather diary in the hope that it could be useful for medical purposes, though he did not chronicle incidents of sickness in the same journal. Robert Hooke's scheme for recording the weather called for compilers to make daily annotations of illnesses in their localities. Christopher Wren requested an annual report on prevailing diseases from the physicians of the society, together with notes on the "difference of operation in medicine according to the weather and the seasons." [9] Thomas Sydenham, a leading London physician and friend of many members of the society (though never himself a fellow), began to note weather conditions in his records of diseases in the early 1670s. Sydenham became an influential advocate of the systematic, empirical approach to the subject. The development of his research showed how themes of the Hippocratic tradition were reshaped in light of the contemporary interest in diurnal changes of weather. He began in the classical Hippocratic mode by identifying seasonal patterns in the occurrence of "fevers" (by which he meant all acute, as opposed to chronic, conditions). Winter typically brought coughs, gout, measles, and smallpox, for example; spring yielded pleurisy and dysentery; summer brought cholera, scarlet fever, and smallpox again; and autumn would see rheumatism and the return of dysentery. Each season had its complement of "tertians" and "quartans," fevers that reached their peaks of intensity every three or four days, respectively. As he began to attend to daily changes of weather, Sydenham realized that the seasonal pattern would be interrupted by sudden alterations in atmospheric conditions, which could usher in diseases that were anomalous for the time of year. He distinguished "stationary fevers," characteristic of a particular season, from the "intercurrent fevers" that might interrupt them when the weather changed dramatically. Thinking along the same lines as Boyle, he speculated that changing weather conditions affected the atmosphere's capacity for absorbing the earthy exhalations that caused disease. [10]

Sydenham bequeathed to his many admirers and followers in the eighteenth century a set of issues for investigation and a technique for confronting them. Many subsequent medical writers were to ponder the relationship between the patterns of the British climate and the diseases that seemed to recur periodically in the population. They were also to try repeatedly to distinguish illnesses typical of a particular season from those that owed their origin to some unexpected alteration in the weather. The principal method of these inquiries, for many decades, was the combination journal of weather and diseases. In 1723, the physician and secretary of the Royal Society James Jurin published his general invitation to the learned world to submit meteorological records, expressing the hope that the project would advance medical knowledge. He received at least eighteen submissions, many by doctors, from as far away as Italy and Scandinavia. [11] A few years later, Francis Clifton recommended recording episodes of disease in the form of a table that would also include the weather. Clifton was a leading advocate for the Hippocratic method in the early 1730s, an editor of the classical texts, the author of a history of medicine, and a leading member of the Royal College of Physicians of London.[12] The prominent London physician and man of letters John Arbuthnot, whose *Essay concerning the Effects of Air on Human Bodies* (1733) was the most widely read theoretical discussion of the atmospheric causes of disease, also called for the compilation of medico-meteorological journals. Hippocrates' plea for such research had not been heeded, complained Arbuthnot, but if it were, "a Piece of Knowledge, almost scientifick, might be founded, not incurious or useless to Mankind."[13]

Jurin, Clifton, and Arbuthnot provided significant encouragement for the Hippocratic project, which began to gain momentum in the third decade of the eighteenth century.[14] Hans Sloane published the journal he had kept in Jamaica back in the 1680s, and gave added authority to the medico-meteorological program through his presidency of the Royal Society from 1727 to 1741.[15] In slightly more obscure circumstances, and apparently independently of the metropolitan medical community, Clifton Wintringham, a physician at the county hospital in York, began in 1715 to record the weather and the diseases he observed in his practice. Wintringham explained the rationale for his enterprise by invoking the classical tradition, naming as inspiration not only Hippocrates and Galen but also the Roman physician Celsus and the philosophers Plato and Lucretius.[16] He continued his journal until 1734, publishing the record as a condensed narrative that surveyed a season at a time. Wintringham had obviously made instrumental measurements quite frequently, but in the published account he commented only on episodes when extremes of temperature or pressure were reached or when a dramatic change occurred. For each season, he gave a list of the diseases he had observed, remarking on the seasonality of certain illnesses and how they seemed to increase or diminish when the weather changed. In footnotes to the text, he groped toward explanations of these phenomena in terms of the physiological impact of changes in the atmosphere. Wintringham clearly understood that these tentative explanations were much less securely grounded than the basic factual data conveyed by his text. The narrative form of the journal was his solution to the problem of organizing these fundamental empirical facts, raw materials—as he hoped—for answering the Hippocratic questions of how diseases changed with the seasons and the more rapid fluctuations of the British weather.[17]

All investigators faced the difficulty of finding a form of writing that would allow these questions to be addressed. In 1741, Roger Pickering presented a plan to the Royal Society for incorporating records of diseases in a tabular weather register. The aim was to find out more about the national propensity to certain illnesses, to guard "against the Disorders, which, as *Islanders*, we are exposed to." [18] Pickering, a country vicar who also wrote an essay on mushrooms in the *Philosophical Transactions*, proposed that the standard columns for temperature, pressure, rainfall, and so on be supplemented with one giving figures for the causes of deaths drawn from the weekly bills of mortality. His scheme has been shown by Andrea Rusnock to mark a significant step toward the compilation of social statistics. [19] But statistical tabulation did not displace narrative description, and most researchers continued to rely on their own encounters with those suffering from diseases rather than on published figures. As Pickering put it, scientific knowledge "must arise from a Variety of Observations, made by different Men of Application and Judgment." [20] Printed tables were often used for recording instrumental readings, but most observers thought it more appropriate to give verbal narratives when it came to recounting their personal experiences with patients.

John Huxham, a physician in practice in Plymouth, was encouraged to start keeping a medico-meteorological journal by Jurin in the mid-1720s. Huxham had interests in natural history and astronomy as well as in many aspects of medicine; he used a barometer and a thermometer supplied by the London instrument maker Francis Hauksbee to begin his weather journal in 1724. A few years later he started to record illnesses, adopting the form of a month-by-month narrative in which he passed from remarks about the weather to comments about the diseases he had encountered. Like Wintringham, he noted extremes of air pressure, temperature, or rainfall, or occasions when a sudden change had occurred. He paid particular attention to the wind's direction and when it altered; he recorded mental conditions—low spirits, melancholy, lunacy—along with the physical ailments that afflicted local people. [21] A very similar approach was adopted by William Hillary, also a physician, who began his record just about the same time, at Ripon in Yorkshire. Hillary kept up his journal from 1726 to 1734, when he moved to Bath. Later, he resumed his inquiries in Barbados, to which he emigrated in the early 1750s. [22] Hillary produced a continuous narrative, punctuated sometimes by months and sometimes by seasons. Though he had evidently used instruments, he gave few readings of thermometer or barometer; but he did note episodes when they changed abruptly. Sudden alterations in the weather were especially remarked when they coincided with the emergence or disappearance of diseases. The episodic narrative form allowed Hillary to recount these incidents as it were historically, suggesting by the construction of his prose a connection between the atmospheric causes and the pathological effects to which they were thought to give rise. He packed into the narrative framework details about symptoms and the effectiveness of particular therapies, and he was able to mention individuals who, because of their peculiar constitutions, were exceptions to the general trends. A typical paragraph from his years in Yorkshire ran as follows:

> The State of the Weather continued much the same till about the middle of *March* 1728, when the Barometer which had risen a little before, now fell again, as did the Thermometer also, and we had great and almost continued Rains for three Weeks, with cold North Winds, and sometimes Snow: Upon which the above-mentioned Inflammatory Diseases, *viz.* Pleurisies, Peripneumonies, Quinseys, and some Rheumatisms, increased both as to the Number of the Sick, and the Violence of their Symptoms; the Pleurisies and Peripneumonies were some of them of the true, and others of the nothous [spurious] Kind; the Pulse was generally low, but very quick and hard, the Pains acute, the Blood very sizy, and cover'd with a thick buff-like Pellicle; the Sick were not relieved without often repeated Bleeding, diluting plentifully with Emollients and Pectorals, and Volatiles added to them. Those who before had suffered much from the Intermittent in the Winter, were most liable to be seized with these Disorders; and as they could not bear the Loss of much Blood, many of them died.[23]

At the end of his record, Hillary added a series of "aphorisms," drawing preliminary—and admittedly speculative—conclusions from his observations. His comments indicate how the Hippocratic program had been reinforced, in the decades before he wrote, by concepts derived from the iatromechanical tradition of the seventeenth century, which understood the workings of the human body in terms of the mechanical operations of fluids and solids. Wintringham, Arbuthnot, and others drew upon the same theoretical vocabulary in their accounts of the effects of atmospheric conditions on health. Hillary suggested, for example, that hot weather relaxed bodily fibers but reduced their elasticity, and that it led to the volatile parts of fluids being excreted in perspiration, leaving the thicker and less mobile parts behind. The result would be obstructed circulation, inflammations, and fevers. Cold, on the other hand, contracted the fibers, reducing the speed of circulation and the release of noxious matter by perspiration. Changes of weather conditions thus removed the causes of certain diseases but tended to give rise to others. Rapid changes of conditions would be hazardous for everyone, particularly for those with more susceptible constitutions. Hillary concluded that, from his observations, "the Reasons will appear, Why temperate, moist and gradually variable Weather is most healthful: Why sudden Changes of the Seasons are, on the contrary, most sickly: And, Whence it is, that each Season, if it observes its common Course, is attended with Disorders peculiar to itself." [24]

Hillary's work showed what a perceptive and thoughtful observer might accomplish working within the eighteenth-century Hippocratic tradition. The new awareness of weather as a quotidian phenomenon was concentrating people's attention on the often rapid changes of conditions that were typical of the national climate. It sometimes seemed that the only constant thing about the British weather was change. This being so, the traditional Hippocratic focus on the seasonality of diseases required some modification in the circumstances. Hillary knew that each season would produce the normal complement of diseases "if it observes its common Course." But in order to determine if the season was typical, it was necessary to keep track of conditions on a shorter timescale, such as the diurnal one of the daily journal. This was the procedure generally adopted among the British devotees of Hippocrates, who then faced the problem of trying to discern a pattern of connections between abrupt changes in the weather and the ups and downs of diseases in individual patients or in the population as a whole. In 1733, the prominent physician George Cheyne commented that the standard cyclical patterns that were supposed to govern the development of diseases, identified by Hippocrates in the calm and settled conditions of the Mediterranean, were hopelessly confused "in this various and inconstant Climate." [25] Forty years later, John Rutty, drawing upon his own long experience in Dublin, agreed that trying to correlate the weather of the British Isles with the progress of diseases was as hopeless as fixing a sundial on a weathercock. [26]

Notwithstanding the difficulties, compilers of medico-meteorological journals continued their efforts. Rutty himself incorporated notes on the prevailing diseases in his record of the weather in Dublin from 1725 to 1766. He claimed to have kept his journal continuously, but he worked it up for publication into a narrative covering a month at a time. He knew the work of Wintringham, Huxham, and others, and concurred with the consensus view that the perennial dampness of the British Isles, with its temperature oscillating seasonally between moderate warmth and moderate cold, produced constant "endemic" ailments, including a fever of "a low putrid kind." [27] Likewise, Rutty agreed that each season had its own complement of illnesses. In addition, temporary epidemics swept across the land, frequently

Gent: Mag: Vol: XXI : completed Dec! 1751.

ÆRE PERE

S. Wale delin?

C. Grignion Sculp.

Begin with Sense, of ev'ry Art the Soul,
Parts answ'ring parts shall slide into a Whole.

Pope.

FIGURE 16 · Frontispiece to the *Gentleman's Magazine* 21
(1751). This plate introduced the volume of the *Gentleman's
Magazine* in which John Fothergill's weather diaries first
appeared. It shows Asclepius having laid down his rod entwined
with a serpent to consult a barometer. Courtesy of Special
Collections, Dimond Library, University of New Hampshire.

coinciding with abrupt changes of weather. Rutty used published accounts from London, Edinburgh, and other locations to map the progress of these. John Fothergill's monthly accounts of the weather and diseases in London, published in the *Gentleman's Magazine* from 1751 to 1754, adopted a similar approach. He made use of his instrumental record to tabulate the extremes of temperature and pressure in each month, and also the greatest change of these variables in any single day. Both factors were significant because, according to Fothergill, "not only a steady course of any kind of weather may produce particular diseases, but likewise very quick transitions from one extreme to another may be equally injurious." [28]

Everyone studying the subject seems to have agreed on what they were looking for: on the one hand, recurring patterns of seasonal ailments, and on the other, the impact of sudden changes in weather conditions. Most authors claimed to have found such patterns among the patients they were acquainted with, but it proved difficult to generalize the conclusions in any satisfactory way. In 1762, Charles Bisset tried to summarize the consensus in his *Essay on the Medical Constitution of Great Britain*, and he found himself harshly criticized for the attempt. Bisset was an Edinburgh-educated physician who had served as an army surgeon in the West Indies and North America and as a military engineer in the Low Countries before settling into private practice in Yorkshire. He began his essay with remarks that reflected the consensus about the British climate, noting that its temperateness and moisture were usually healthy features, while "the native prevailing diseases in this Island [are] in general generated, and excited, by the frequent changes of the weather peculiar to Great Britain." He went on to catalogue the ailments typical of each season, noting that anomalous illnesses were always liable to be introduced by unseasonable weather and sudden alterations in the air. Resorting to iatromechanical reasoning of the kind that was typical of the Hippocratic writers, Bisset suggested that the effects of moisture and moderate cold strengthened the fibers and overall fabric of the body, so that "the natives of Great Britain, in general, are bigger bodied, broader chested, and more robust, than those of most other countries." Vulnerable though they were to colds, rheumatism, and certain fevers, the British people could be assured that their climate was basically a healthy one, and that "epidemic diseases of great malignity are much greater strangers in this island, than in most countries on the continent." [29]

Bisset probably did not anticipate criticism for these fairly anodyne variations on the providential interpretation of the British climate. Like many other authors, he was telling his co-nationals that although they were vulnerable to certain airborne illnesses, things could have been much worse, and indeed they were in other countries. He was fiercely taken to task, however, by a writer for the periodical *The Critical Review*, edited by the novelist and physician Tobias Smollett. The reviewer (probably Smollett himself) claimed that Bisset had overemphasized the seasonal character of British ailments, applying a system that "might be proper in a country that enjoys a regular succession of seasons, with sure and settled tracts of weather," but that was inappropriate for "Great Britain, where the transitions from one extreme to another are so sudden and irregular." Bisset was also said to have erred in his optimistic view of the healthy qualities of British rain. The reviewer maintained that "if we may trust to our own observation, a rainy winter is, of all winters, the most unwholesome; nay, it is proverbially so."[30] Bisset was stung by the criticism, especially by complaints about the tedium of his prose style and its contamination by crude Scottish expressions; he felt obliged to defend himself by bringing out a pamphlet in response.[31] But, considered from a distance, the dispute reveals not so much fundamental differences of outlook as the basic assumptions of the Hippocratic tradition, which were common to Bisset and his critical reviewer. Both held that the atmospheric environment was a primary cause of prevailing diseases; both attempted to understand this in iatromechanical terms of the influence of heat and humidity on bodily fibers and fluids. Both believed that seasonal variations and the more abrupt changes of weather had effects on the occurrence of diseases, though they tended to give different degrees of emphasis to the two factors. And both operated within a providential framework in which the British climate was generally thought to be good for people—although, as physicians, they were obviously well aware that sickness and death were unavoidable, and to that extent they acknowledged that all could not always be for the best.

These were the leading themes of the Hippocratic tradition as it took root in eighteenth-century Britain. Newly conscious of the quotidian character of their weather, British investigators from Sydenham onward realized that the traditional Hippocratic emphasis on the seasonality of diseases required some modification. Most observers recognized seasonal patterns in the occurrences of particular ailments, but they also pointed to the impact of more rapid changes of atmospheric conditions—the daily fluctuations typical of the British weather. To trace the effects of these on patterns of sickness required constant monitoring, and the compilation of medico-meteorological journals seemed the appropriate way to approach the problem. With some variations of form, these continued to be written throughout the century. Thomas Short, though he sometimes expressed skepticism about the value of these records and turned to the methods of social statistics later in his career, kept his own journal in Sheffield for nearly thirty years.[32] In the 1770s, Dr. John Bayly in Chichester, Sussex, was adding copious medical notes to his painstaking monthly records of the weather and expressing confidence that eventually, "ye Causes of Diseases may be rationally deduced from ye manifest Qualities of ye Air." [33] At the end of the century, yet another provincial doctor, Thomas Hughes from Stroud in Gloucestershire, was recording occasions when "many cough at church" on Sundays, gauging the health of his neighbors as meticulously as he measured air temperature, pressure, and humidity with his meteorological instruments.[34] Hughes was working in a tradition of inquiry that was, by his day, already more than a century old. Like his many precursors, he was seeking ways to grasp the variables of weather and disease, convinced that Hippocrates had been right to suppose that the climatic environment had an important bearing on the health of a population. For British investigators, the Hippocratic perspective suggested that changes in weather conditions would cause particular illnesses to emerge. Prolonged investigation and direct personal observation were thought to be necessary to uncover these coincidences. Notwithstanding the efforts of many dedicated inquirers, however, the temporal scales of atmospheric and pathological events proved impossible to reconcile, and the connections between weather and disease continued to elude definitive specification.

Aerial Sensitivity and Social Change

In *The Spectator* of 25 July 1712, Joseph Addison offered his readers an account of "a Sett of merry Fellows, who are passing their Summer together in the Country." The group was said to be residing in a substantial house, which, along with apartments for all of the company, contained an infirmary "for the Reception of such of them as are any way Indisposed, or out of Humour." As the story unfolds, the members succumb one by one to bad temper, melancholy, or some other condition that manifests itself in antisocial behavior, and are dispatched to the infirmary. The narrator speculates in a Hippocratic vein about how immoderate diet or inclement weather might have led to this epidemic of indisposition. In this setting, readers are introduced to the figure of the "human barometer." One of the company having announced to the rest that "he knew by a Pain in his Shoulder that we should have some Rain, the President ordered him to be removed, and placed as a Weather-glass in the Apartment above-mentioned." [35] Addison's satirical sketch evoked the seventeenth-century weather glass, which, as we saw in the previous chapter, was often likened to the human body in its responses to changes in the air. By the beginning of the eighteenth century, the barometer had begun to take over this role. It displayed the physiological effects of the atmosphere and reminded people that the human body itself could be considered a kind of instrument. People who manifested a heightened sensitivity to the air's qualities, or said that they could tell when the weather was going to change, became known as human weather glasses or human barometers.

When the barometer was still quite new, researchers insisted that it displayed properties of the air that were physiologically significant, even though they sometimes disagreed about what those properties were. In 1673, Boyle wrote that he was "prone to suspect" that alterations in atmospheric pressure could affect human health.[36] Martin Lister, who had a theory of the barometer quite different from Boyle's, thought that the device indicated the effects of the air on the bodily humors.[37] John Smith and Richard Neve, early writers on the instrument for the general public, held that it showed the atmosphere's influence on the body. As Smith put it, "The lower the *Quicksilver* descends, the more listless and out of order Men's Bodies are." When the mercury level was high, on the other hand, according to Neve, "Men's Bodies are then found to be more Brisk and Lively."[38] John Arbuthnot described the medical consequences of changes in air pressure in his influential essay of 1733. He wrote that he had "observ'd very sensible Effects of sudden falls of the Mercury in the Barometer in tender People, and all the Symptoms they would have felt by the Exsuction of so much Air in an Air-Pump." Susceptible individuals, according to Arbuthnot, experienced "lypothymies" during sudden drops in air pressure, undergoing convulsions like the mice and birds Boyle had sacrificed in his air pump.[39] In 1750, Thomas Short summarized what was by then a medical consensus. In conditions of high pressure, Short wrote, "we find ourselves brisk and lively, from the greater Velocity of the Blood, and fuller and juster Discharge of all natural and necessary Secretions and Evacuations." Excessively high pressure, however, posed risks to health, tending to bring on such illnesses as pleurisy, pneumonia, and hot fevers. Low air pressure was equally hazardous, causing a dangerous diminution of circulation and perspiration that could induce hysteria, nervous disorders, and putrid fevers.[40]

Concern about barometric pressure did not exclude awareness of the dangers posed by other factors in the aerial environment. In fact, atmospheric pressure diminished as a preoccupation of medical writers later in the century, as discussion of the perils of tropical climates increased. In the tropics, heat and humidity seemed to be the most pressing dangers to the health of British soldiers and settlers. But medical opinion agreed that the changes revealed by the barometer were physiologically significant, especially if they were rapid and of large magnitude. Thus, barometers, along with thermometers, hygrometers, and wind gauges, became part of the instrumental armory of researchers recording the fluctuations of weather and diseases. Wintringham used the barometer and thermometer while compiling his record in York for almost two decades. Huxham used both instruments and a hygrometer for his *Observations on the Air and Epidemic Diseases* (1739). Hillary observed the barometer for nine years in Yorkshire and then for a further six in Barbados. John Phelps's satirical poem *The Human Barometer* (1743) reflected the general view of the time that high pressure was good for circulation and vivacity, while low pressure led to languor and melancholy. Use of the barometer had made people aware of a new dimension of climatic sensitivity, showing them that their bodily health was vulnerable to the environment in a previously unsuspected way.

Phelps's poem also spoke to widespread moral concerns about atmospheric sensibility. The notion that large numbers of individuals were succumbing to environmental illnesses was a worrying one for many commentators. It suggested that the circumstances of modern life might be responsible, either by aggravating the noxious qualities of the air or by weakening people's resistance to its effects. Increased susceptibility was often thought to flow from moral failings, such as intemperance or indulgence in luxury or soft living. Phelps explained in the prose prologue to his poem that moral and physical health alike depended on the mind's control over the impulses of the body. The union of body and soul, he explained, is "so intimate tho' inexplicable" that each was profoundly influenced by the state of the other. Virtue demanded "the proper Exercise of the rational Faculties, to maintain a regular and watchful Government" over the impulses of the body, so that they did not prejudice one's health.[41] Illness, according to Phelps and others, could be a sign of moral weakness. Those whose health suffered from the effects of the air were suspected of being partly responsible for their own condition, having given in to their passions or weakened their constitutions through intemperance.

Phelps sketched some of the unfortunate social consequences that could follow from these lapses of self-control. He compared two instances of extreme mental derangement: the madness of the inmates of Bedlam, London's notorious lunatic asylum, and the religious fanaticism of followers of the Methodist preacher George Whitefield, who was renowned for his ability to mesmerize his audiences. Whitefield was presented as preaching just outside the walls of Bedlam, evoking from his hearers the same kind of frenzied behavior found among the inmates of the institution.

There mounted on his Tripod *Whitefield* stands,
Silence and Awe canonick Garb commands,
With Arm extended see he apes *Saint Paul*,
And counts his own an Apostolick Call,
Gesture and Voice betray the heated Brain
In Groans his Converts echo back again,
And Souls impress'd with Thoughts of Grace, or Sin,
Expectorate their Sense in solemn Din.
These of enthusiastick Transports boast,
But are to Argument and Reason lost.[42]

With the mention of "enthusiastick Transports," Phelps reminded his readers of the associations surrounding the highly charged term *enthusiasm*. Since the mid-seventeenth century, "enthusiasts" had been portrayed as in the sway of their own dangerous passions, claiming religious authority for what were actually mental disturbances caused by vapors rising into the brain. Henry More wrote in 1662 that enthusiasts were inspired by "nothing else but that *Flatulency* which is in the *Melancholy* complexion, and rises out of the *Hypochondriacal* humour upon some occasional heat."[43] In this way, religious fervor was classified as a kind of sickness; fanaticism—with its destabilizing social consequences, so painfully evident in the conflicts of the seventeenth century—was "medicalized."[44] In Phelps's poem, the seventeenth-century enthusiast was reincarnated as the eighteenth-century human barometer. Climatic susceptibility emerged as a kind of surrender to the passions, closely allied to melancholia, hypochondria, and enthusiasm, and raising similar worries about its potential antisocial consequences.

The concern about these conditions in the early eighteenth century was heightened by trends that seemed to be encouraging people to be more sensitive to their surroundings. What has been called the culture of sensibility was manifested in new attitudes and behavior that were becoming prevalent among the British middle classes by the middle of the century. Polite manners were supposed to be accompanied by refined feelings, shown by an individual's aesthetic responses and sympathy for other people and animals. Philosophers, following the lead of the third earl of Shaftesbury, argued that sympathy was an expression of an inherent "moral sense," which was the basis for telling right from wrong and acting accordingly. Good taste was also said to derive from inherent feelings; people were thought to have a capacity to recognize beauty when they encountered it. Such sentiments were natural, but they also had to be cultivated. Politeness was

FIGURE 17 · William Blake, "Air—on Cloudy Doubts &
Reasoning Cares" (1793). Blake's sketch evokes the
traditional association of melancholy with cloudy weather.
© Copyright Fitzwilliam Museum, University of
Cambridge. Reproduced by permission.

identified with the refinement of one's inherent feelings as guides to what
was good in morals, manners, and the arts. With the spread of these attitudes, more considerate behavior came to be expected of individuals, and
more emotional responses to literature, art, and music became common. As
G. J. Barker-Benfield has shown, these developments had important effects
on the relations between the sexes in the period. Middle-class men tended
to abandon cruel sports and rough behavior, turning to the cultivation of
manners and emotions previously associated with women. The figure of
the man who had gone too far in this direction and had become foppish or
effeminate was a stock character for satire in novels and drama.[45]

Medical writers worried that the culture of sensibility was bringing
with it a tendency to a certain kind of pathology. Sensitivity, it was feared,

could readily become *too* developed; refinement or delicacy could tip over into debility or illness. *The English Malady* (1733), by the charismatic and eccentric doctor George Cheyne, was a harbinger of this concern. Cheyne, a Scottish physician who made his fortune in practice in London and Bath, regarded nervous sensitivity both as an attribute of intellectual and social refinement and as a disposition to ill health.[46] Enhanced sensibility could all too easily lead to nervous disorders, as those of refined manners or cultivated intellect frequently found to their cost. Some of the causes of this were general to the British nation, according to Cheyne. He pointed to the climatic situation of the island at the edge of an ocean, with its moist air and variable weather. More important were the social and economic trends that had increased luxury and encouraged self-indulgence. Cheyne condemned such modern habits as eating food flavored with rich sauces, riding in smooth coaches, and living in crowded cities. His contemporaries were putting their health at serious risk by indulging in this kind of pleasure. They should take a lesson from the ancient Greeks, who had found that "in Proportion as they . . . distinguished themselves from other Nations by their Politeness and Refinement, they sunk into *Effeminacy, Luxury*, and *Diseases*, and begun to study *Physick*, to remedy those Evils which their Luxury and Laziness had brought upon them."[47]

Cheyne's anxiety about the diseases of modern life was very influential. He highlighted for his contemporaries concerns about the medical damage wrought by the ease and affluence of civilized society. The English malady—which Cheyne identified particularly with nervous disorders such as hypochondria, hysteria, and melancholia—came to be seen as an expression of the country's prosperity and the excessive sensitivity it had spawned. Like other medical writers, Cheyne was confident that the remedy for these ills was temperance: a moderation of consumption in all respects, a return to natural foods, vigorous exercise, and clean living. He identified temperance with virtue and was convinced that virtue would be rewarded by good health and happiness. Although in this respect he shared the providentialist assumptions of his contemporaries, he also gave expression to an underlying anxiety. He suggested not only that the British climate was less than perfectly conducive to the wellness of the population, but that specific aspects of the modern lifestyle were making things worse. He pointed to the dangers of living in a congested city like London, where the smoke of fires and candles, human breath and perspiration, the ordure of animals and people, and the effluvia of graveyards, slaughterhouses, and dunghills were "more than sufficient to putrify, poison, and infect the Air for twenty Miles round."[48] The risks posed by such development should

not be ignored. As far as Cheyne was concerned, those who looked to seasonal fluctuations to explain outbreaks of disease were missing the much
more serious impact of contemporary social changes.[49] The Hippocratic
model, rooted in the placid climate of the classical Mediterranean, was not
really applicable to modern Britain, where nervous disorders were more
prevalent because people's sensitivity had been increased by the contemporary environment.

Cheyne did not view the atmosphere as the only source of modern maladies, but his concerns fed into the enterprise that was searching for the
aerial origins of disease. The idea that modern life was increasing people's
pathological susceptibility qualified some of the prevailing complacency
about the benefits of the national climate. On the one hand, the optimistic
view was that the British population was blessed with just the right degree
of nervous sensibility, with positive consequences for their civilization and
even their political liberties. The physician William Falconer argued that
their natural sensitivity made them good at friendship, respectful of the
rights of women, sympathetic to the plight of the less fortunate, strong in
the defense of their independence, and fond of animals. The British, he
asserted, owed their social sentiments to their climatic situation. As inhabitants of the temperate zone, they were able to tame their feelings and direct them to the ends of civilized society. Their emotional stability allowed
them to experience the benefits of collective life while rejecting authoritar-
ian tyranny. Had they lived further north, they would have been brutal and
antisocial, like the inhabitants of the polar regions; further south, the hot
passions of the tropics prevailed, which were equally destructive of social
harmony.[50] As the Scottish historian William Robertson wrote, it was only
in the temperate region of the globe that mankind "possesses a superior
extent of capacity, greater fertility of imagination, more enterprising cour-
age, and a sensibility of heart which gives birth to passions, not only ardent,
but persevering."[51]

On the other hand, however, there were those who worried that this
emotional stability was slipping away and that sensibility was running out
of control. Samuel Johnson seems to have thought that to admit that the
weather could affect one's health or mood was to submit the powers of the
mind to the sway of the passions. In *The Idler* for 24 June 1758, he complained that "surely nothing is more reproachful to a being endowed with
reason, than to resign its powers to the influence of the air. . . . To call upon
the sun for peace and gaiety, or deprecate the clouds lest sorrow should
overwhelm us, is the cowardice of idleness, and the idolatry of folly."[52]
Johnson thought his contemporaries were far too willing to indulge their

atmospheric susceptibilities; he repeatedly told James Boswell in conversation that it was foolish for people to say that the weather affected their feelings. He also denied that his massive consumption of tea influenced him emotionally, though some doctors blamed the beverage for increasing people's vulnerability to atmospheric ailments.[53] For Johnson, keeping control of his reasoning powers was a point of pride, and the fear of losing them was a source of great anxiety. He suffered severe and repeated bouts of depression, dreaded that he was going mad, and prayed frantically that he would not lose his mind. He had no time for the fashionable notion that a melancholy disposition was a sign of artistic genius; for him, it was simply a source of acute suffering, compounded by the dread that it might foretell a complete loss of mental control.[54] Given these preoccupations, it is not surprising that Johnson regarded with stern disapproval the tendency among his contemporaries to submit themselves to the qualities of the air.

For Johnson, the issue of climatic sensitivity was a highly personal one. Other writers of a conservative or moralistic inclination emphasized how it was symptomatic of undesirable trends in their society. They saw many of the new fashions of the time as examples of indulgence and intemperance that would surely weaken people's resistance to airborne maladies. Wintringham remarked on the unhealthy quality of urban air and the debilitating effects of drinking tea and coffee, "so much in Use among the Ladies." [55] Short and Fothergill shared his view that excessive consumption of tea could make people more vulnerable to atmospheric diseases. Hillary, after emigrating to Barbados in the 1750s, castigated the European residents there who were so in thrall to fashion that they wore clothing much too heavy for the hot climate and contracted a variety of diseases as a result. He lamented that "Fashion and Custom are two prevailing Things, which inslave the greatest Part of Mankind." The predilection of European women for dancing also attracted his reproach, though he acknowledged that "most of the Ladies are so excessive fond of it, that say what I will they will dance on." [56]

Much of the criticism was directed at women's behavior, a perennial target of moralistic censure under the cover of medical advice. Arbuthnot identified as particularly unhealthy the unventilated rooms in which "People of Fashion pass a great deal of their time." "Ladies and other tender People," he noted, suffered the effects of air "tainted very much with the Steams of Animals and Candles." [57] New fashions and manners among women were always liable to attract men's disapproval, since they challenged male authority. But the yoking of women with other "tender people" and "people of fashion" suggests that the targets of criticism were more general and the issues at stake more specific to this context. Climatic

susceptibility was thought to be symptomatic of an overall process of softening or effeminacy, which was believed to be affecting men as well as
women in eighteenth-century Britain. The critics were fastening on general developments in the society around them: the accelerating cycles of
fashion, the increase in material affluence or luxury, the consumption of
new foodstuffs and drinks, and the tendency to refined manners, emotional
expression, and aesthetic sensitivity. All seemed to reflect the attenuation
of masculine identity in the face of a feminization of character and mores.[58]
And all were represented as increasing the population's risk of succumbing
to diseases, including those originating in the air.

Thomas Short's work in the 1750s and 1760s shows how medical discourse was influenced by these moralistic concerns. Around the middle of
the century, he began to compile statistics to try to answer questions about
the health effects of climate and the local environment. This culminated in
A Comparative History of the Increase and Decrease of Mankind (1767), a
pioneering work of social statistics, in which Short used the bills of mortality to deduce which places in England and other countries had the healthiest soil, water, air, and other conditions of life.[59] Overall, he concluded that
healthy locations were dry, on mountains or rocky soil, while unhealthy
ones were wet, low-lying, and swampy. In addition to these physical factors, his analysis devoted considerable attention to social conditions like
legislation and prevailing moral standards. He was reflecting the debate
(to be discussed in the next chapter) in which David Hume and others had
responded to the ideas of the Abbé Du Bos and the Baron de Montesquieu,
distinguishing the "moral" from the "physical" causes of longevity and
population growth. When he contemplated the notoriously high mortality
rates in London, Short turned to moral causes to explain them, pointing out
the frequency of prostitution and intemperance in the capital. Promiscuity,
according to Short, decreased human fertility, as did excessive consumption of liquor. Firm measures were required by legislators and magistrates
to suppress this kind of vice and promote virtue, in order to ensure the
health of the population.[60]

Striking this moralistic note, Short echoed the works of his medical
contemporaries and predecessors, including Cheyne, Arbuthnot, Wintringham, and Hillary, who viewed susceptibility to atmospheric diseases
as symptomatic of moral weakness. He shared their concern about the
fashions of modern life that had increased the prevalence of this aerial
sensibility. He also, however, endorsed more concrete programs of environmental improvement, which were being pursued in the second half of
the eighteenth century. Short stressed that it was possible to take practical

steps to "mend the air." He mentioned improving the ventilation of houses, draining marshy areas near settlements, removing human and animal excrement, and other measures. Initiatives of this sort were commonly endorsed by enlightened reformers and were implemented to some extent in the latter part of the century. They were to take the project of atmospheric medicine, which had emerged from the Hippocratic revival of the late seventeenth century, in a new—and at times quite controversial—direction.

The Politics of Atmospheric Reform

The revival of the Hippocratic tradition posed problems for the providential interpretation of the British climate. Medical practitioners tried to determine if the familiar fluctuations of the island's weather were healthy or unhealthy for the population. Did they confer the benefits of a stimulating variation, as was suggested by some authors, or did they expose people to a constant succession of new diseases, as classical tradition might lead one to expect? More fundamentally, the question of climatic susceptibility raised pressing moral concerns surrounding the relations between the physical environment and the progress of civilization. Some writers developed the providential line, arguing that the British people owed to their climate a moderate and socially beneficial degree of sentimental feeling, while others suggested that the population was being put at risk of airborne and other diseases by a sensibility accentuated by the conditions of modern life.

Among these conditions were the circumstances of urban existence, increasingly registered as hazardous to the health of those who lived in London and other large towns. Many commentators expressed anxiety about the unhealthiness of urban spaces, the air polluted by the effluvia of so many people and animals, their carcasses, rubbish, and bodily wastes. Rutty's complaint about the air in Dublin in the early 1770s was typical. He pointed to

> the fogginess from the smoke when there is no wind to dissipate it, the dirtiness of our streets, which is so great that one is frequently in danger of being up to the knees in crossing them, the putrid animal effluvia exhaling from Charnel-houses and dunghills in the middle of the city and in several of the avenues, and dead animals, dogs and cats and the excrements of living ones, butcher's garbage and blood, and burying grounds likewise in the middle of the city, where the earth, in the graves, is frequently so loose and the bodies so near the surface of the ground, that the scent has been noxious in a hot summer.[61]

The Hippocratic tradition of medical thought made these atmospheric dangers seem particularly pressing. Members of the polite classes believed their enhanced sensibility was making them more vulnerable to the effects of such contamination. At the same time, they prided themselves on a refined sense of smell that brought them warning of the perils. Projects to improve the quality of the air unfolded against this background, using foul odors to identify sources of unhealthy airs and intervening in the environment to remove them. The early part of the century saw attempts to bring healthy air into the places where people lived or worked, initially in small-scale projects to ventilate buildings, mines, and ships. Within a few decades, more ambitious proposals were being made. The urban atmosphere was to be comprehensively improved by programs of environmental amelioration and social engineering. Stagnant water was to be drained, cemeteries cleared from urban centers, sewage and wastes removed, and streets widened to allow for the circulation of air.[62] Discoveries about pneumatic chemistry and the mechanism of respiration encouraged the advocates of these proposals. Researchers believed they could distinguish "good" from "bad" air by chemical tests; they assumed a single scale of aerial quality, identifying the healthiness of the air with the absence of foul smells. The goodness of the air was thus associated with moral virtue, reflecting the values invested in hopes for atmospheric improvement. The advocates of improvement held that good air was naturally conducive to good health, while bad air was the product of combustion, putrefaction, and death. They believed that progress would come from exploiting the benevolence of providence to offset the effects of atmospheric corruption, supplying people with the air that was good for them while removing that which was bad.[63]

A prominent pioneer of pneumatic chemistry and atmospheric improvement in Britain was Stephen Hales, curate of Teddington in Middlesex from 1709 until his death in 1761. Hales was known for his investigations of animal and plant physiology and for developing methods to collect quantities of air released by chemical reactions or the processes of life. He did not chemically differentiate the kinds of air he collected; instead, he focused on the physical property of expansiveness that they all shared. This, he believed, was the key to the role of airy substances in the providential economy of nature, a universal system sustained by the balance between repulsive and attractive forces among particles of matter. Hales believed providence must have provided means to restore the natural expansiveness of air after the vitiation caused by combustion or respiration, though he did not know what these mechanisms were. He therefore stressed the importance of human intervention to bring good air to where it was needed in order to replenish that which had been exhausted. This was the motivation for his design of a machine for ventilating ships' holds, prisons, and hospitals. The need for fresh air was particularly urgent, according to Hales, where the atmosphere had been tainted by combustion, putrefying matter, diseased persons, or the bodies of the dead.[64]

The general conception of an economy of good and bad air, and the association of bad air with putrefaction and disease, continued to feature in the development of pneumatic chemistry by Hales's successors. Sir John Pringle was particularly influential in encouraging work on atmospheric improvement in the three decades after 1750. Having resigned a medical professorship at Edinburgh, Pringle served for a while with the British army in Flanders in the late 1740s. His *Observations on the Diseases of the Army* (1752) drew attention to the dangers of establishing field hospitals and camps near the putrid air of marshes. In line with the Hippocratic tradition, he traced the onset of epidemics to meteorological conditions, but he also advocated vigorous intervention to change the air surrounding sites of habitation. To deal with scurvy—then a scourge of the Royal Navy as it tried to extend its influence across the world's oceans—Pringle recommended ventilation of ships and the consumption of certain "antiseptic" substances, such as fermenting vegetables and the "fixed air" (carbon dioxide) recently studied by the Glasgow chemistry professor Joseph Black.[65]

Pringle also supported the work of the Dissenting minister Joseph Priestley, whose pneumatic researches of the early 1770s yielded dramatic new resources for studying varieties of air and putting them to use. In 1772, Priestley—then a minister at the Mill Hill Chapel in Leeds—published a pamphlet that gave directions for dissolving fixed air in water. What we know as carbonated or soda water was seen by its inventor as an artificial replacement for the naturally aerated waters of certain mineral springs, which had long been regarded as good for the health. Apparatus for making the impregnated water was soon widely distributed; these machines, sometimes called gasogenes, appeared by the thousands in middle-class dining rooms before the end of the decade. Commercial manufacture of carbonated waters followed very soon thereafter, and supplies were taken on sea voyages to combat scurvy. Priestley regarded the health-giving properties of carbonated waters as God's gift to humanity, a providential reward for ingenuity and rational inquiry. Also in 1772, he published a celebrated paper in the *Philosophical Transactions* that described how air vitiated by combustion or respiration could have its "virtue" restored by allowing plants to grow in it or agitating it over water. Commenting on the process two years later in the first volume of his *Experiments and*

Observations on Different Kinds of Air, Priestley remarked that methods for restoring the goodness of air were part of the providential design, essential to the economy of nature as a whole.[66] He was confident that all putrid, respired, and noxious airs could be restored by these means. Thus, the role of plants in the order of nature was revealed, as was a hitherto unsuspected function of storms at sea. As Pringle put it, in a speech to the Royal Society in Priestley's honor, the agitation of the oceans served "to bury in the deep those putrid and pestilential effluvia which the vegetables upon the face of the Earth have been insufficient to consume."[67]

A further contribution of Priestley's 1772 paper was the "nitrous air test," which from this point assumed a critical importance in pneumatic chemistry as a way of assessing the goodness of air. In this procedure, a sample was mixed with what Priestley called "nitrous air," and the resulting product was observed to diminish in volume as part of it was absorbed by water. The degree of diminution was said to be in proportion to the purity of the sample, since it could be correlated with other measures of aerial purity, such as the ability to support combustion or respiration. For Priestley, the nitrous air test was a convenient gauge of aerial virtue; it was immediately seized upon by researchers looking for ways to measure the quality of the air in different places. Marsilio Landriani in Milan and Felice Fontana in Florence developed apparatus to perform the test and took them to urban and rural locations in northern Italy, making measurements of the healthiness of the air and its seasonal variations. Landriani's term *eudiometer* (from the Greek for "measure of good air") became the general name for this class of instrument. Designs varied from Priestley's own simple collection of tubes with a basin of water to the elaborate ivory and crystal presentation piece given by Landriani to Count Firmian, the counselor of state in Habsburg Lombardy.[68]

For a while in the late 1770s and early 1780s, it seemed that eudiometry would answer exactly the needs of aerial reform. Instruments were taken on field trips in England, Italy, France, and elsewhere, and they were even manufactured commercially. The York physician William White published a eudiometrical survey of the atmosphere of his city in the *Philosophical Transactions* in 1778. On a visit to London the following year, Fontana assessed the quality of the air in the streets and at the top of the dome of St. Paul's Cathedral. In 1780, the Dutch physician Jan Ingenhousz reported to Pringle on eudiometrical measurements made in the course of a voyage from London to the Netherlands. Usually, the results of these expeditions confirmed what was expected: the air was found to be purer out at sea than near to the coast, and decidedly impure in the vicinity of marshes or in crowded city streets. But although the eudiometer could often be calibrated by reference to the sense of smell, long used to distinguish salubrious from insalubrious air, it turned out to be impossible to translate sensory impressions reliably into quantified measurements. The field descended into a series of fierce disputes among practitioners, each claiming a procedure that would yield replicable results while denouncing those of the others. Fontana attempted to introduce discipline into methods of measurement, and his system attracted some followers in England; but there was never a consensus that reproducible accuracy had been achieved. By the mid-1780s, it was generally doubted that nitrous air eudiometry would be able to provide the precise quantification of the healthiness of air that its early advocates had hoped for.[69]

The underlying assumptions of the eudiometrical program nonetheless persisted among those whom Priestley influenced. Kinds of air were to be arranged in a single scale of virtue, corresponding to their suitability for respiration and their general healthiness. Priestley understood this scale in terms of the theory of phlogiston, believed by chemists to be the principle of inflammability. Foul air was thought to be heavily contaminated with the phlogiston released by burning bodies and respiring people and animals. The less phlogiston in it, the better the air. In 1774, when Priestley produced a kind of air that could support respiration for even longer than normal atmospheric air, he named it "dephlogisticated air" and assumed that its high degree of respirability would make it especially healthy. Breathing the air, he reported a light and easy feeling in his chest; he speculated that "in time, this pure air may become a fashionable article in luxury."[70] The suggestion that this air (subsequently renamed "oxygen") could enhance health was soon investigated by physicians who were already using fixed air therapeutically. A group of doctors in the major provincial towns reported to Priestley about their experiments in treating patients with various gases. William Hey (in Leeds), Thomas Percival (in Manchester), Matthew Dobson (in Liverpool), John Haygarth (in Chester), and William Falconer (in Bath) were among the practitioners who adopted pneumatic therapy and became leading advocates for its effectiveness. Many of them had existing interests in the quality of the local atmosphere and its effect on the health of the population. Priestley's discoveries presented them with resources that were adaptable to local reform projects and to the therapeutic practices of enlightened doctors.[71]

While the findings of pneumatic chemistry gave medical practitioners new therapies for individual patients, reformers did not lose sight of the wider social agenda of the Hippocratic tradition. In fact, it was given a radical new orientation in the last decades of the century. Priestley had encouraged many people to think about the role of scientific knowledge in general enlightenment and social reform. Aerial improvement could be seen as part of the overall progress of society, which for Priestley manifested the providential destiny of humankind. For him, technology or "art" did no more than exploit the God-given capabilities of nature; human progress could be expected to follow the path pointed out by divine guidance. Priestley believed that limitless progress in humans' intellectual, moral, and material condition would be the consequence of the spread of scientific knowledge. To those who shared his perspective, new knowledge of the medical benefits of gases appeared as a sign of the advance of society toward enlightenment. Obstructing this process — for the time being — were the agents of corruption: political tyranny, religious superstition, human folly. But Priestley was confident that these shadows would vanish as the light of knowledge dawned throughout society. Nothing could long resist the power of truth to make people healthy and free.[72]

Adam Walker, a Manchester schoolmaster who turned himself into a successful public lecturer in London and other cities, was one of the first publicists for Priestley's pneumatic researches after the author himself. He started to include them in his lectures in York in the early 1770s, and was later given apparatus by Priestley to use in his displays. Walker seems to have shared Priestley's views about the providential character of scientific discoveries and the role of education in public enlightenment. He lauded the rational knowledge of God that came from the study of nature, while castigating superstition and political tyranny. Writing in 1778 of the problem of bad air in large cities, Walker presented the issue in its moral context:

> It cannot too often or too forcibly be inculcated, how necessary to Health is the breathing of good air. When religious tyranny huddled its absurd votaries together near churches and monasteries; plagues, pestilences and famine announced the outrage unheard; 'twas the immediate finger of God, in the language of ungrateful and ignorant fatalists. . . . It may seem strange that in this age of philosophy and enlarged sentiment, we should run into similar error; but so it is; tho' we have opened our streets, pulled down our signs, and made sewers for every thing that may contaminate the air; a Court can seduce the active and needy with its employments, the rich and idle with its pleasures, and all with its Luxuries, Douceurs, and Fashions. Hence . . . our minds lose their relish for simplicity and nature; and

even the Lungs accommodate themselves to a thick and putrid air, so as to be even offended by that of the Country.... It remains but for the philosopher to *moderate* the evil if possible, by his researches.[73]

Bemoaning the unhealthy consequences of luxury and fashion, Walker echoed a standard trope of the Hippocratic tradition. The artificiality of modern life had long been seen as exposing people to diseases, including those originating in the air. But Walker gave the lament a specifically political tinge by mentioning "religious tyranny" and linking it to the corruption of the "Court." He thereby associated the seductions of luxury and fashion with the superstitions of medieval Catholicism, an archetype of the obstacles that stood in the path of enlightenment. He also sounded a chord common in eighteenth-century oppositional rhetoric, which frequently denounced the royal court and the government ministers thought to be its lackeys. By the 1770s, this language was being taken up by the popular forces beginning to demand a greater degree of participation in the political process. Walker was aligning the campaign for aerial improvement with that against the corruption of an unrepresentative parliament and the established church, an orientation that reflected Priestley's own view of the radical political implications of his scientific discoveries.

In the 1790s, as partisan divisions deepened in British society in response to the French Revolution, and as Priestley was driven into exile in America following an attack on his Birmingham house by a reactionary mob, the program of pneumatic medicine assumed a strongly political coloration in the work of Thomas Beddoes, who took up the torch of Priestley's campaign. His medical training, his knowledge of pneumatic chemistry, and his political radicalism made him an appropriate inheritor of Priestley's legacy, though ultimately his reputation suffered even more severely from conservative scorn. In some respects, Beddoes's medical views descended from Cheyne's. He worried about the debilitating effects of nervous sensibility on people's health. Like many before him, Beddoes was convinced that the population was being softened and its resistance to illnesses weakened by the conditions of modern life, including fashionable clothing and people's indulgence in music and frivolous literature. He believed that increased nervous sensitivity led to such respiratory ailments as rheumatism, asthma, and consumption. One of Beddoes's therapeutic experiments took the pastoral, anti-urban theme of the Hippocratic tradition in a startling new direction. As a treatment for consumption, he recommended that patients sleep in cowsheds, in close proximity to the fumes and excrement of the animals. Ladies of high fashion and developed sensibility could profit from this treatment, he assured his readers, provided they could steel themselves to the stench and the indignity.[74]

Beddoes began offering gaseous therapy to patients at his Pneumatic Institution, founded in Bristol in 1797. He presented it as part of a program to reform medical practice by making patients more responsible for their own health and reducing the authority of professional practitioners. His aims were consistent with ideals of comprehensive social enlightenment and were supported by sponsors who included many of Priestley's old friends, such as the doctors William Withering and Erasmus Darwin. They also, however, aroused opposition from the established medical profession and from a British government increasingly paranoid about political subversion. At one point, Beddoes was forced to admit, "I know well that my politics have been very injurious to the airs." [75] This opposition was fed by the dramatic discovery of the properties of nitrous oxide, introduced in 1799 by Beddoes's assistant, the young chemist Humphry Davy. Breathing this new gas produced effects of euphoria and giddiness—apparently like intoxication, but without the subsequent hangover. Davy and Beddoes made the gas available to a circle of their acquaintances, including the poets Robert Southey and Samuel Taylor Coleridge. The effects were widely reported and were used by conservative writers as a pretext to ridicule the whole program of pneumatic medicine. The Tory periodical *Anti-Jacobin Review* published two poetic satires targeting Beddoes's therapeutics along with the radical ideas of his friend Darwin. "The Pneumatic Revellers: An Eclogue," published in 1800, portrayed Beddoes and his colleagues as wild enthusiasts, using the gases discovered by the satanic Priestley to enjoy orgies of intoxication and sexual license. They were said to have been carried away by the force of their own imaginations, convinced that the new gases would usher in a utopian age of universal enlightenment in which mankind would "feed on *Oxygene*, and never die." [76]

Sadly, from Beddoes's point of view, the satire was far too close to the mark for comfort. He had indeed written of nitrous oxide and other gases as material agents of a possible universal enlightenment. He speculated with Davy about the chances that a "sublime chemistry" would make available to everyone the means to perpetuate pleasure and remove pain. But in a more pessimistic mood, he worried that "we might even prepare a happier æra for mankind, and yet earn from the mass of our contemporaries nothing better than the title of enthusiasts." [77] That was in fact just what happened. Beddoes and Darwin never recovered their reputations from the opprobrium and ridicule heaped upon them at this point. Both died within a few years, their contributions to science and human welfare largely discredited. The younger men associated with them, like Davy and Coleridge, tacked to the right politically and tried to put their involvement in the nitrous oxide fiasco behind them. But the incident continued to be brought up occasionally by critics and satirists. In Thomas Love Peacock's novel *Nightmare Abbey* (1818), a character who seems to be based on Coleridge is identified by a young lady with the words, "You are a philosopher ... and a lover of liberty. You are the author of a treatise called 'Philosophical Gas; or, a Project for the General Illumination of the Human Mind.' " [78]

The reduction of pneumatic medicine to a farcical scheme for the diffusion of "philosophical gas" was an attempt to denigrate the enlightened ideals that had found expression in the project. In the decades of political conflict that began in the 1790s, an intense struggle occurred to define the character of the Enlightenment as a whole. As a period of European history, the Enlightenment ended at the end of the eighteenth century; as an intellectual outlook with ambitions for freedom and progress, its longer-term survival was precisely what was at issue in the debates of the time. To this day, our understanding of the movement remains marked by the controversy that swirled around it as its historical moment came to a close. One aspect of this controversy was the fate of pneumatic medicine and atmospheric reform. Pneumatics as a form of individual therapy was disgraced, tainted by association with the despised philosophy of materialism, which Priestley had publicly articulated. This line of criticism revived the earlier strictures of Samuel Johnson against submitting the powers of the mind to the influences of the air. Pneumatic therapy was said to have attempted to use material influences on the body to fulfill the desires and aspirations of humanity; it had promised spiritual improvement through manipulation of the passions. This allowed its advocates to be portrayed as the latest incarnations of the seventeenth-century enthusiasts. Notwithstanding their declared faith in reason, they were accused of having surrendered their judgment to the intoxicating effects of artificial airs.

Johnson had prefigured this satirical characterization in an episode of his novel, *The History of Rasselas, Prince of Abyssinia* (1759). The hero of the tale, Imlac, tells of his encounter with an astronomer who believes that prolonged and deep study has given him the power to control the weather throughout the world: "The clouds, at my call, have poured their waters, and the Nile has overflowed at my command." Imlac concludes, however, that the astronomer is mad, deluded by an overactive imagination into believing that godlike powers are in his hands. He ascribes the man's madness to scholarly melancholy and isolation from the refreshing diversions of

society: "No man will be found in whose mind airy notions do not sometimes tyrannize, and force him to hope or fear beyond the limits of sober
probability. All power of fancy over reason is a degree of insanity."[79] This
was exactly the diagnosis of the pneumatic reformers by conservative critics forty years later. Surrendering to their own "airy notions," they were
said to have allowed their ambitious imaginations to trump their powers
of reason. They had thus brought ridicule upon themselves by trying to
spread enlightenment by gaseous means.

Although individual pneumatic therapy was largely discredited at the
end of the eighteenth century, ideas about the atmospheric origins of disease did not die out; nor did attempts to address them by environmental improvement. In this respect, the legacy of the Hippocratic tradition
was part of the inheritance of the Enlightenment as a whole.[80] "Mending
the air" continued to be a priority of activists in the emerging domain of
public health. Bad air remained the distinguishing hazard of unhealthy
places, associated with stagnant water, rotting waste, sewage, and corpses.
Air was regarded as the primary vehicle by which putrefaction gave rise
to diseases, and so irrigation was urged in order to carry away putrefying
matter before it could infect the atmosphere. Thus, projects for drainage
and sewage removal were prompted by foul odors and judged by the criterion of breathable air. As the great Victorian sanitary reformer Edwin
Chadwick put it, "All smell is disease."[81] Furthermore, although the great
sanitation projects of the nineteenth century unfolded in a very different
social and political context, they remained rooted in conceptions of providence that descended from the Enlightenment, as Christopher Hamlin has
pointed out. The Victorian reformers retained the notion of a single scale
of aerial virtue; they assumed that the goal of human improvement was to
restore the natural goodness of air as the key to health.[82] These assumptions
underlay Chadwick's *Report on the Sanitary Condition of the Labouring
Population* (1842), which in many respects carried forward the aspirations
of enlightened atmospheric reformers. Notwithstanding the political crisis
that had surrounded it at the end of the previous century, the ambition of
changing the air in order to improve human health lived on, grounded still
in an enlightened confidence in the providential goodness of nature.

This conviction had been shared by all sides in the eighteenth-century
controversies. Cheyne, Short, and others castigated their contemporaries for
their self-indulgence in luxury and fashion, for having sacrificed their natural robustness of health to modern comforts. They assumed that a return
to temperance and self-control would restore the nation's health, because
nature contained within itself the remedies for human disorders. If one ad-

justed one's life to live in a natural manner, one's health would be repaired by the natural goodness of the air. Priestley, Walker, and Beddoes, on the other hand, believed that the air in many places had been damaged by human misconduct. But again, the remedy lay within nature itself. Restoring the natural goodness of the air—by artificial means, if necessary—would allow people to recover their good health. Both individual therapeutics and large-scale projects for social reform were guided by this conviction. The common assumption of enlightened thinkers, whether conservative moralists or progressive reformers, was that the air was naturally good for human life and that providence had provided remedies for situations in which its natural virtue had been corrupted.

Climate and Civilization

Nothing that depends on the social state, is so unalterably fixed, but that it will change and vary with the degradation or improvement of the human race. And hence, while the nature of man remains unaltered, the state of society is perpetually changing, and the men of one age and country, in many respects appear different from those of another. And as men themselves are more or less improved, every thing that constitutes a part of the social state, will bear a different appearance among different nations, and in the same nation in different circumstances, and in different periods of time.

SAMUEL WILLIAMS · *The Natural and Civil History of Vermont*

IN THE PREVIOUS CHAPTERS, we have considered how people's understanding of the weather in eighteenth-century Britain reflected changes in their society and culture. As they tried to make sense of their experiences of weather, they were made aware of the cultural transformation of their time—incomplete though it was when measured against the aspirations for comprehensive enlightenment. They came to recognize how hopes for the triumph of reason and social progress were constrained by the physical limits of human nature and the historical inheritance of attitudes and beliefs. Thus, we saw how "impolite" weather phenomena raised fears among the elite that unenlightened patterns of behavior would return, how "superstition" seemed to survive in connection with calendar lore and weather prediction, how even new instruments sometimes seemed to be treated like magical oracles, and how the influence of weather on health demonstrated the vulnerability of human reason to bodily passions. In all these cases, experiences of the weather mirrored the circumstances of people at the time, as beings with physical bodies, situated in a specific historical context. How the weather was perceived was, in that sense, reflexive of the experience of enlightenment itself, which was always accompanied by an awareness of its incompleteness.

Relations between the physical environment and human culture were also extensively discussed by eighteenth-century intellectuals. In particular, they debated the relationship between climate and the progress of civilization. This manifested another dimension of Enlightenment reflexivity, namely, the consciousness of how nature and human life mutually shape one another. In eighteenth-century discussions, climate stood for nature itself; it signified the physical circumstances of existence in their bearing on human life. The key point is that nature was not regarded as an external force acting upon human beings from the outside. People were regarded as unavoidably *part* of nature, bound to it by the "human nature" that was thought to constitute their essence. There was a wide range of opinions as to the makeup of human nature—about the importance of its material component in relation to its spiritual or intellectual component, for example. But all eighteenth-century thinkers agreed that humans had an essential nature, grounded in their physical being and the circumstances of the world around them. They set out to derive knowledge of morals, society, and history by specifying this nature, assigning it a normative force in determining how people should live and an epistemological function as the key to understanding them. As Roger Smith has put it, "Nature itself thus set the conditions, Enlightenment writers argued, which made experience and history possible, and the language of nature set the terms in which man was to be understood."[1]

Furthermore, when writers of the time invoked nature, they did not think of it as set against culture or society. They did not operate with the conceptual dichotomies that would oppose these things to one another. Indeed, the concepts of "culture" and "society," as they are familiar in the modern human sciences, developed only in the nineteenth century.[2] Enlightenment thinkers assumed that what we call society and culture were manifestations—of one kind or another—of human nature. When they speculated about a "state of nature," they engaged in a fictional exercise designed to strip away the artificial elements and get back to what was natural, which was thought by many to provide a key to how people ought to live. Whether the state of nature currently existed in some part of the world, or had existed at a specific time in the past, was somewhat beside the point. The reason it was invoked was to draw out essential features of human nature. The conclusions were much debated, but what was not contested was that, in all the varieties of civilized and uncivilized life, human nature was being expressed. This meant that material connections between human beings and their physical environment were always of interest. Intellectuals debated the magnitude of climatic influences, their relative importance in comparison with the other forces shaping human life, and how strongly they were expressed at different stages of the development of civilization. Throughout these debates, it was assumed that human beings were — to some degree or other — subject to the circumstances of their physical environment as a condition of their existence as creatures with material bodies.

The question of the bearing of climate on the development of human civilization was a particularly urgent one for Europeans who settled in other parts of the world. British colonists in North America, the West Indies, and India appreciated immediately that the climates in those places were very different from that prevailing at home. They set about investigating local conditions, using the techniques of systematic recording and instrumental measurement that had been used to chart the weather in Britain. They generated data for a series of comparisons, favorable and unfavorable, with the homeland. Physicians and surgeons serving with the British armed forces or practicing in the colonies carried the Hippocratic preoccupation with the effect of climate on human health into their new situation. They tried to assess the influence of heat and humidity, of soils and winds, of marshes and forests, on the settlers. While recognizing the different conditions, they also tended to perpetuate the concerns of British commentators with, for example, the dangers of luxury and fashion. They pondered the limits the local climate might set to the development of colonial society, and they also worried that inappropriate habits brought from home might make settlers especially vulnerable to airborne diseases.

The British colonies in North America presented a kind of laboratory for assessing how climate affected civilization. Early in the eighteenth century, American settlers began to adopt British methods of recording and measuring their weather. As they built up a picture of atmospheric conditions, they debated the influence of these conditions on prospects for social development. While the physical environment was seen in some respects as a threat to American society, the colonists also believed they were taming it by extending settlements, cutting down forests, and bringing the wilderness under cultivation. Nature, including the climate, could be "civilized" by these means. After the revolutionary break with Britain, Americans developed their sense of themselves as an independent people. They often spoke of nature as an ally in the cause of independence and an asset that would support the growth of their civilization. Increasingly, they took pride in their climate, which they believed was being reshaped to serve the needs of the new nation. Like the British before them, the Americans came to value their weather as a national resource, one that contributed to their destiny. But whereas the British climate was thought suitable to an oceanic island with an important maritime role, the Americans saw theirs as fitting for a continental power destined to bring a vast territory under the sway of civilization.

The Enlightenment Debate on Climate

The eighteenth-century debate on climate and civilization unfolded within the framework of attempts to grasp human nature and understand its different manifestations in laws, manners, and customs. It was fed by the empirical inquiry into weather and its bearing on human health, which, as we have seen, had taken root in Britain in the early decades of the century. Although it drew upon ancient traditions of thought, the ensuing debate reflected distinctly eighteenth-century concerns with the history of civilization and its relations to nature. Thinkers approaching the issue often invoked certain dichotomies—for example, between the natural and the artificial, or between the mind and the body. But these dichotomies proved hard to sustain consistently and tended to become unstable. It was concluded, for instance, that even very artificial modes of life laid people open to the influence of the air on their health. Similarly, it was thought that the workings of even a highly refined mind could be subverted by atmospheric forces acting on the passions. Civilization, it seemed, did not permit people to escape the influences of their climate, any more than they could evade their human nature.

When the ancient writers talked of "climates," they referred to a notional division of the world into zones of latitude: frigid, temperate, and tropical. Each zone was supposed to be inhabited by people whose characteristics derived from the prevailing heat or cold of their atmosphere. Medical writers in the Hippocratic tradition had gone further than this, exploring human sensitivity to properties of the physical environment—including the air, waters, and soil—in specific places. As geographical knowledge expanded in the early-modern period, climate remained an important conceptual resource for coping with the diversity of human mores and institutions. Enlightened intellectuals used the idea to reconcile the variability of humankind with the concept of an underlying human nature. But climate did not hold this privileged place in intellectual discourse for very long. By the early nineteenth century, attempts to stretch it to explain the whole range of human diversity came to seem artificial and unrealistic. As new disciplines developed, including biology and sociology, people came to understand the environment and its bearing on human life in new ways.[3] At the same time, many human differences came to be regarded as attributes of the different "races," supposedly rooted in the biological inheritance of the various strains of humanity. From this later standpoint, eighteenth-century ideas about climatic influences on cultural development seemed crude and confused. Historians who try to recover enlightened thinking on these questions confront a characteristically fluid situation, prior to the emergence of the disciplines that form the core of the modern social sciences. Eighteenth-century thinkers were pressing their intellectual inheritance to the limit. They were deploying the idea of climate to try to encompass the range of human social life at a time of considerable cultural change, when the very categories of thought they used to reflect on their societies were also being transformed.

The first eighteenth-century work to use climate as a key to comparative history was the *Réflexions critiques sur la poésie et sur la peinture* (*Critical Reflections on Poetry and Painting*), published in 1719 by Jean-Baptiste Du Bos and translated into English in 1748. Du Bos argued forcefully for the operation of physical causes in the rise of artistic creativity. He professed to see no other reason why certain settings—classical Athens, say, or Renaissance Italy—should have produced such extraordinary outbursts of cultural expression, while strenuous efforts by patrons and governments to reproduce them elsewhere had failed. Du Bos was vague on exactly what he meant by physical causes, and he offered little explanation of how they operated. He repeatedly used metaphors of natural growth, suggesting that the arts sprang up from the ground like well-nurtured crops. Although he could not say precisely what they were, he insisted that the qualities of the air were of crucial importance in determining the mental character of different nations: "The difference between the air of two countries is imperceptible to our senses, and out of the reach of any of our instruments; for we know it only by its effects."[4] Though he sometimes talked of the ancient climatic zones of the globe, implying that he saw temperature as the most important atmospheric variable, Du Bos's language mostly invoked a more general "nature" as the prime stimulus for intellectual vigor. The mind was said to be particularly susceptible to the influences of nature, conveyed through the qualities of the air.

Du Bos pointed the way to naturalistic accounts of cultural development that would encompass climatic influences, but he did not connect this with medical theories of the air's effects on the human body. That link was made by John Arbuthnot, in his *Essay concerning the Effects of Air on Human Bodies* (1733). Arbuthnot added to his Hippocratic discussion of the weather's influence on diseases speculations about how it also affected the characters of different nations. He mentioned a suggestion by Hippocrates, which was echoed by Du Bos, that Asians were disposed to accept despotism because of moral weakness caused by the hot climate. This became a common prejudice among European writers in the eighteenth century and was given renewed currency by Montesquieu. Those who lived in more bracing and variable climates were supposedly stimulated to more industrious and courageous activities. Arbuthnot thought that "Mathematicians, Philosophers, and Mechanicks" would tend to arise in a nation with this kind of climate, whereas "Painters, Statuaries, Architects, and Poets, which, besides the Rules of Art, demand Imagination" would come predominantly from warmer places.[5] As we have seen, the notion that climatic variability stimulated mental alertness was taken up by other British writers. Passions such as the imagination, on the other hand, tended to be seen as products of the balmier climes of the Mediterranean. Arbuthnot even suggested that variations in language might be ascribed to climatic differences. "The serrated close way of Speaking of Northern Nations, may be owing to their Reluctance to open their Mouth wide in cold Air," he suggested, "which must make their Language abound in Consonants; whereas from a contrary Cause, the Inhabitants of warmer Climates opening their Mouths, must form a softer Language, abounding in Vowels."[6]

The ideas of Du Bos, Arbuthnot, and others were taken up by Charles Louis de Secondat, Baron de Montesquieu, generally held to have been the most important writer on climate and civilization in the eighteenth century. While the originality of Montesquieu's thinking on this topic has probably been overestimated, his work was highly significant for integrating climatic influences into an account of social and political structures. He relied quite heavily on previous writers for his conception of how atmospheric properties affected human physiology, but he went much further, by situating climate in a comprehensive comparison of different societies that was widely read and debated. Although many subsequent thinkers disputed the role he had ascribed to climate, they were obliged to engage with Montesquieu's arguments. This kept ideas of climatic influences alive in all discussions of social development and its causes.

For Montesquieu, temperature was the atmospheric variable with the most important physiological effects. His *Essay on Causes Affecting Minds and Characters* (written in 1736–43, but unpublished until the nineteenth century) discussed the effects of heat and cold on the nerve fibers, which in turn shaped the characters of different nations. Northerners were said to be sluggish but sound in their mental judgments, while southerners were quicker-witted but more subject to the passions.[7] At the same time, Montesquieu acknowledged other environmental influences, such as winds, qualities of the soil, and local foods. These were considered additional "physical" causes of specific mental characteristics. In his magnum opus, *Esprit des lois* (*Spirit of the Laws*; 1748), all of these causes found their place in the analysis of different structures of law and government. A few of Montesquieu's lines became notorious: his statement that "you must flay a Muscovite alive to make him feel"; his comparison of the audiences' reactions to operas in England and Italy; his experiment of freezing a sheep's tongue to observe the contraction of its nerve endings.[8] These images laid him open to criticism and even to satire, but they do not adequately represent his argument as a whole. Already in the *Essay*, he had listed "moral" causes that were counterparts to physical ones in forming the "general character" of a people. They included education, laws, religion, customs, and manners.

In the *Spirit of the Laws*, Montesquieu explored the relationship between the physical and moral factors lying behind national characteristics. He showed how social and political formations could be analyzed by isolating their essential features and tracing them to their underlying causes. Although systematic in its overall organization, the work proceeded by assembling aphorisms rather than by articulating a connected argument, so it was not easy for readers to extract a single point of view on this question. Many thought Montesquieu had given too much emphasis to climatic causes, but he also often stressed their subordination to moral ones. Where climate tended to weaken the moral strength of the population, he declared, legislators should act forcefully to counter its effects. Whereas "Nature and climate almost alone dominate savages," the societies of more civilized nations, such as the Chinese and Japanese, were governed by manners and laws.[9] By and large, climate made itself felt on the body, and Montesquieu supposed that—at least in more enlightened societies—the body would be subordinated to the rule of the mind. Thus, climate remained an important factor in the study of human society, but it was expected to be eclipsed in importance by manners and customs as civilization advanced.[10] As one might anticipate with an enlightened thinker, education in particular was ascribed great importance in enabling people to free themselves from the constraints of their physical environment. Nonetheless, by placing climatic influences on a level to be compared with those of customs and law, Montesquieu had suggested that physical and moral causes were somehow equivalent. This kept the physical environment in play as a factor in the Enlightenment debate about social development, even though other writers disagreed sharply about how important it was.

A significant criticism of Montesquieu's theory was that it failed to articulate an account of social progress. Although he repeatedly stated that climatic influences could be subordinated by education and enlightened legislation, he did not spell out how societies developed toward a situation where these factors would prevail. His countryman Anne Robert Jacques Turgot initiated a line of thought among French intellectuals in which the scale of progress was used as a key for differentiating human societies, and in which climatic causes of variability were given much less emphasis.[11] In Britain, and particularly in Scotland, where questions of progress were also central to social theory, Montesquieu's ideas were subjected to fairly rigorous critique. The tone was set by David Hume, whose essay "Of National Characters" appeared in the same year as the *Spirit of the Laws*, and may have been framed as a response to it.[12] Hume systematically demolished the idea that climate alone could account for the variations between national characteristics, giving no less than nine instances of counterexamples: nations or peoples that enjoyed the same climate but were markedly different in characteristics, or places where the climate had remained constant but the attributes of the inhabitants had changed. He did not mention Montesquieu by name, and it is possible that his argument was aimed not at him but at the much cruder climatic determinism of Du Bos. Hume concluded that climate could not explain the differences among populations of the temperate zone, such as those of southern and northern Europe. He did, however, concede that there was reason to believe that peoples who lived "beyond the polar circles or between the tropics, are inferior to the rest of the species, and are incapable of the higher attainments of the human mind."[13]

Hume's concession was a revealing one, not just because it exposed a strain of prejudice that was lamentably common among European writers on human diversity, but also because it touched upon a difficulty with his argument. Skeptical as he was about the purported influence of climate, Hume nonetheless resorted to "nature" to account for what he considered undeniable differences in the cultural achievements of different peoples. In a footnote to his essay that has since become notorious, he asserted that black Africans were "naturally inferior to whites," since there "scarcely was a civilized nation of that complexion." He concluded, "Such a uniform and constant difference could not happen, in so many countries and ages, if nature had not made an original distinction between these breeds of men." [14] So while Hume claimed that the influence of natural causes could be transcended by "civilized" nations, he was still willing to invoke them to explain the supposed backwardness of other peoples. Evidently, the physical environment could not be entirely discounted in the analysis of human development.

In principle, Hume's position was not all that dissimilar from Montesquieu's. Both expected morals, customs, and laws in enlightened nations to modify the influences of the physical environment; both resorted to nature to account for the state of uncivilized peoples. Furthermore, Hume's account of life in civilized societies left the door open to a degree of climatic influence that he was reluctant to admit explicitly. In his essay, he defined "moral," as opposed to physical, causes as "all circumstances which are fitted to work on the mind as motives or reasons, and which render a peculiar set of manners habitual to us." [15] The emphasis in his philosophy on habit, customs, and the passions, and the limited role he ascribed to reason in motivating human action, assigned prime importance to factors that had often been seen as susceptible to environmental influences. When he wrote in his essay that national characters were due to "a sympathy or contagion of manners," he named emotional attributes frequently viewed as subject to the forces of the atmosphere.[16] As we saw in the previous chapter, for many of Hume's contemporaries, "sympathy" was among the human characteristics most susceptible to the changing qualities of the air. Hume, it seems, had not squashed the argument for climatic influences on human life as conclusively as he claimed.

Hume's essay—both in its categorical assertions and in its ambiguities—set the tone for other Scottish writers in their response to the climatic theory in the 1760s and 1770s. On the one hand, Montesquieu was readily criticized by those who wanted to downplay the role of climate, often crudely reduced to the effects of heat and cold. On the other hand, physical nature could not be entirely excluded, especially when it came to discussions of the ways of life of "primitive" peoples or the role of sentiment and the passions in what were thought of as more advanced societies. The leading Scottish writers on comparative or "conjectural" history—John Millar, Adam Ferguson, and Henry Home (Lord Kames)—developed the four-stage theory of progress that saw all human societies as passing through a sequence of phases distinguished by their mode of subsistence. They acknowledged that climate was decisive for the earlier stages, when people lived in small bands of hunters and gatherers or as nomads, but they routinely objected to Montesquieu's suggestion that it determined significant differences between nations at the more advanced stages, once settled agriculture and commerce had been established. At the same time, the possibility that climate could interfere with the process of social development itself could not be completely ruled out. Kames declared that a hot climate would prevent society from developing beyond the hunter-gatherer stage.[17] Millar wrote that differences among the English, Irish, and Scots, who shared essentially the same weather, showed that "national character depends very little upon the immediate operation of climate." But he also admitted that too little was known of what effects it might have.[18] Ferguson followed Hume in allowing that temperate Europe might have been climatically destined to lead the way in civilization. He wrote that European primacy manifested "either a distinguished advantage of situation, or a natural superiority of mind." [19]

Europe's apparent advantage in mounting the ladder of progress was one thing that made it difficult to discount the influence of the physical environment entirely. It was hard to see what else could explain how social progress had begun, even if one held that climatic influences had diminished as the process continued. Around the turn of the nineteenth century, theories of the human "races" began to be developed, which assigned different levels of intellectual capability to different populations. Then, European superiority was ascribed to the biological inheritance of the white race.[20] Kames pointed the way to this development by introducing the idea of "polygenesis," the notion that different strains of humanity had originated separately. But Kames also ascribed a role to climate in causing the "degeneration" of animal and human types in the New World, and he believed temperate conditions had aided European progress in civilization.[21] Most writers of the time did not regard racial markers as fixed or fundamental aspects of identity, and even skin color was thought to change under the influence of climatic conditions. Thus, climate was an obvious factor to turn to in order to account for European ascendancy. As the Scottish historian William Robertson put it, mankind "has uniformly attained the greatest perfection of which his nature is capable, in the temperate regions of the globe." [22]

Robertson's *History of America* (1777) reflected the Scottish writers' preoccupation with progress and their attempts to reconcile it with the role of climate. Only the first volume of the projected work was published, dealing with the European discovery of the American continent and the Spanish conquests. Robertson intended to resume the project with an account of the North American settlements after the war between the colonists and the British crown was resolved, but he never did so. The part of the history that did appear featured environmental forces quite prominently. Robertson argued that they affected human vigor and sensitivity. In the New World, he asserted, "the principle of life seems to have been less active and vigorous . . . than in the ancient continent."[23] The weakness of the forces of nature touched the native people as well as the wildlife. According to Robertson, American natives were feeble in their bodily constitution and lacking in the facial hair that was a sign of manliness, sensibility, and sexual passion. He endorsed the conclusion of the French writer Cornelius de Pauw that "under the influence of an unkindly climate, which checks and enervates the principle of life, man never attained in America the perfection which belongs to his nature, but remained an animal of an inferior order, defective in the vigour of his bodily frame, and destitute of sensibility, as well as of force, in the operations of his mind."[24] Europeans, by contrast, had benefited from a temperate climate that induced both vigor and sensibility, qualities that found expression in their triumphant achievements in war, commerce, literature, and the arts. The benefits conferred by their climate had enabled the European nations to conquer the New World and subdue its native peoples. But according to Robertson, progress had also allowed them to modify their own climate and begin to change that of America itself.[25] As we shall see later in this chapter, he was not alone in believing that European settlers were taming the American climate by clearing forests and cultivating the land. Advanced societies were thought to be capable of taking charge of their climatic circumstances and civilizing the nature to which primitive peoples remained subject.

Robertson did not consistently analyze the relationship between climate and progress, but he mentioned the natural environment at a number of points in the course of his narrative, suggesting that it might hinder or encourage social development. He also declared that human progress always follows the same pattern, echoing the theory of his Scottish contemporaries. This seemed to make climate the accelerator or brake on the rate at which a society traveled the path of progress. Advanced nations were said to owe their emotional stability and refined social feelings to the influence of their temperate circumstances. But at the same time, they apparently had the ability to tame their climates and to direct the expression of sexual passion by moral legislation. Robertson was obviously conscious that climatic theory had its limits as a tool of historical explanation. He acknowledged that climate was "more powerful than . . . any other natural cause," and he understood the lure of trying to reduce human behavior to laws of nature. But he concluded that "the operations of men are so complex, that we must not attribute the form which they assume, to the force of a single principle or cause." What he was sometimes inclined to call "the law of climate" could not be applied "without many exceptions." [26]

Robertson's work on America sketched one approach to the problem of integrating the natural environment into a history of the progress of civilization. [27] Addressing the issue more systematically, Robertson's compatriot James Dunbar, a lecturer in moral philosophy at King's College Aberdeen, argued that the impact of natural causes diminished as society improved. Dunbar admitted the influence of the physical environment on society as a whole — on its agricultural methods and on the health of the population. But he denied that this influence extended to the rational mind. Notwithstanding the "mysterious influence" said to operate on the mind from the body, he insisted that these forces could be overcome by the development of intellectual capacities by individuals and governments. Like Robertson, he believed that climate itself could be brought under human dominion, being increasingly subjected to rational improvement as the arts of civilization progressed. [28]

Dunbar seems to have been trying to dispel the worrying moral issues surrounding atmospheric susceptibility and the influence of the passions over individual behavior. In the face of such anxieties, he asserted the autonomy of the individual mind and civilization's power to subdue the forces of nature. Dunbar and his Aberdeen colleagues were uniformly hostile to materialism, which they associated with Hume's religious skepticism. [29] Hume, of course, had ostensibly rejected the influence of climatic forces, but the encouragement his philosophy was thought to give to materialism made it particularly important to try to demarcate between mental processes and the powers of the environment — difficult though it often was to do so. Not everyone was as sanguine as Dunbar that the progress of civilization would enhance the ability of human reason to keep the passions in check. As we have already seen, medical writers continued to insist on human vulnerability to atmospheric ailments, even in supposedly advanced countries. And as Dunbar was completing his *Essays on the History of Mankind* (1780), two Scottish-educated physicians were reasserting the argument that the qualities of the air had a substantial bearing on mental character and intellectual abilities. In 1780, Alexander Wilson produced his work *Some Observations Relative to the Influence of Climate on Vegetable and Animal Bodies*. Wilson had a medical degree from Edinburgh, where he had studied under the renowned clinical teacher William Cullen. His argument led off from what he took to be a consensus view among scholars that inhabitants of the tropical and polar zones were incapable of achieving civilization. He mentioned Montesquieu's famous experiment with the sheep's tongue, but resisted what seemed to be one of its implications: that natives of the polar regions would be hardier than people from temperate ones. According to Wilson, people from frigid or torrid climes had the same physical and moral weaknesses. In these cases, climatic factors had to be invoked to explain their backwardness; it was only in temperate countries that moral causes could be expected to significantly improve the well-being of the people.[30]

A more ambitious articulation of the climatic argument was given by William Falconer, another former student of Cullen's. Falconer had moved into medical practice in Bath and developed an interest in issues of public health. He published *Remarks on the Influence of Climate . . . on the Disposition and Temper . . . of Mankind* in 1781, a book hailed by the twentieth-century scholar Clarence Glacken as "the most remarkable in its scope and tone" of all works on climate and civilization in the eighteenth century.[31] In this six-part work, Falconer tried to chart the influence of the whole range of climatic factors recognized by the Hippocratic tradition: weather, physical geography, diet, and customs. He also broadened the analysis to embrace the demographic knowledge that was emerging from contemporary studies of population. Finally, he attempted to integrate these factors with discussion of the progress of societies up the four-stage scale mapped by the Scottish philosophers. Falconer insisted that the action of environmental forces on the body was "by sympathy communicated to the mind"; the rational intellect could not be insulated from such forces. Emotions such as love, friendship, and social sentiment were highly subject to climatic influences, he claimed, a fact that explained the fortitude of northerners and the indulgence and effeminacy of southerners. In this respect, inhabitants of temperate regions were just as subject to their environment as those living at the poles or the tropics. Falconer allowed that the English tendency to high rates of suicide, remarked upon by Montesquieu and others, was "a disorder of the climate." People who lived in the temperate zone had refined but also inconstant manners. Their fickleness and independence of mind made them willing to experiment with social innovations, and hence allowed their societies to make progress. Unsurprisingly, it was in England that Falconer saw climatic forces converging in a positive direction. The English, more than any other people, he claimed, "possess a great thirst after knowledge, and desire of improvement."[32]

The works of Wilson and Falconer show that the climatic argument had not by any means suffered a fatal blow at Hume's hands. Hume had exposed the cruder attempts to differentiate between nations on the grounds of temperature alone, exemplified by the work of Du Bos. But his argument could not entirely dispose of the possibility that climate had some bearing, at least at certain stages of social development. His successors among Scottish philosophers conceded environmental influences on the body, but tried to hold the line against admitting their action on the mind. Historians struggled to integrate climatic forces into their accounts of social progress, formulated under the influence of the four-stage model. And as we have seen, medical writers—apparently ignoring the reservations of the moral philosophers who feared that such an approach pointed the way to materialism—reasserted the importance of climate three decades after Montesquieu first brought it to widespread attention.

The situation in the 1780s foreshadowed, in some respects, the subsequent fate of climatic accounts of the development of civilization. In the following decades, biological theories of racial differences took over some of the work of explaining why some cultures had risen to advanced stages of civilization while others had not. The category of "race" was broadly conceived to include mental as well as physical characteristics of human beings. Understood in these terms, race came to be regarded as an essential component of inherited identity. The new theories drew upon much accumulated prejudice, especially concerning Africans and Native Americans, as the derogatory remarks of Hume and others testify. In this respect, there was a degree of continuity with Europeans' habitual condescension to other peoples. But race as a theoretical formation was premised on the basic immutability of personal identity, and hence challenged climatic accounts of diversity with their assumptions of the plasticity of human character.

Nonetheless, environmental explanations of human attributes did not entirely die out at the end of the eighteenth century. Instead, they took somewhat different directions in the social and the natural sciences, a dichotomy prefigured by the division between Dunbar and the medical writers Wilson and Falconer. Dunbar resisted the idea that the air could directly affect the individual mind, but he acknowledged an influence of climate on society as a whole. Environmental forces became something like a "social fact," apparent at the level of the collectivity but not at the level of the single individual.[33] This indicated how social analysis was to separate from medico-biological thinking, which, in the work of the medical writers who were contemporary with Dunbar, continued to posit the human body's dependence on its material surroundings. In the nineteenth century, environment or "milieu" became an important theme in sociology, though detached from ideas of weather or climate. It was invoked in connection with a specifically social ontology to explain aspects of collective behavior and attitudes. Biology, on the other hand, assumed the duty of explaining how physical nature affected individual organisms' development. This was a feature of evolutionary theories of organic development, which began to emerge in the 1820s after surviving for decades in the shadowy underground to which fears of materialism had confined them. Nineteenth-century sociology and biology, then, built in different ways on the legacy of eighteenth-century climatic theories. From a retrospective point of view, intellectuals of the earlier period seem to have held several pairs of themes in tense alignment: the individual and the collective, the path of progress and its deviations, plasticity and immutability of character, the laws of nature and historical narrative. These productive—though ultimately unstable—categories of thought provided the framework in which the project of understanding the development of civilization unfolded, and in which the concept of climate proved so irresistible to enlightened thinkers.

Medicine and the Colonial Situation

The interventions by Wilson and Falconer in the debate on climate and civilization in the 1780s are a reminder that medical men—as well as philosophers and historians—were concerned about the influence of the weather on human life. In the previous chapter, we looked at how this concern was expressed in Britain, where the climate was blamed for many medical problems, notwithstanding its generally favorable reputation as an asset to the nation. Even the homely British weather was thought to pose certain risks to health, especially to those whose constitutions were inherently weak or who had made themselves vulnerable by intemperance. The turn of the seasons brought a regular cycle of complaints—colds, coughs, catarrhs, rheumatism, and various fevers—and more rapid changes in the weather could also cause outbreaks of illness. When British people explored and settled in other parts of the world, they faced even more serious threats. Movement to an unfamiliar climate was generally held to open individuals to the risk of a whole range of virulent, and frequently fatal, ailments. The diseases known today as malaria, cholera, typhoid, and others were usually classed as kinds of "fever" by the eighteenth-century doctors who struggled to understand and treat them. Medical practitioners were convinced that the afflictions had their origins in the environment in which Europeans settled. In line with the Hippocratic perspectives already deployed at home, they focused on the physical situation in which the colonists lived, including the qualities of the air. What came to be called "tropical medicine" began with British doctors applying their homegrown ways of thinking to the alien climates in which their countrymen had settled.[34]

In most of these locations, including India, the West Indies, and North America, heat and humidity were reckoned the most dangerous atmospheric qualities. Their hazards could be accentuated by such geographical features as marshes and forests, and by unwise choices of clothing and habits. Hans Sloane, who accompanied the Duke of Albemarle to his posting in Jamaica in the late 1680s, discussed these issues in the published account of his voyage. Sloane wrote at length of the topography, botany, and human geography of the island. He detailed its climate with its heat, rainfall, and refreshing breezes. And he gave extensive descriptions of the illnesses he had treated among the colonists. He identified temperance as crucial to determining a patient's chances of survival. A truly temperate individual might live to be a centenarian in such a climate, he claimed. But one who indulged in excessive eating or drinking, or in the debaucheries of "venery," would surely succumb quickly to the prevailing diseases. Temperance was important not only because—as physicians since antiquity had asserted—it strengthened the constitution, but also because it aided settlers' adaptation to local conditions. The less stress placed on the constitution, the more readily it could adjust to the climate. Sloane advised British colonists in Jamaica to abandon European fashions in clothing and to mimic the manners of native people and African slaves.[35]

All of these themes were echoed in the works of subsequent medical writers. They recognized that tropical climates posed specific health hazards for settlers, due to their stark deviation from the conditions of the British homeland. Heat and humidity were the most obvious differences, noticed by everyone, and they formed the starting point of attempts to trace the physiological causes of tropical diseases. Charles Bisset, who served as a military surgeon in the West Indies and North America in the early 1740s, wrote about the health risks for settlers on the Caribbean islands. Heat tended to rarefy the blood, he claimed, but it also promoted perspiration, which was healthy. The real danger came when the air was humid as well as hot. Then the fibers of the body could become dangerously "relaxed," leading to fevers and diarrhea. Newly arrived settlers were particularly vulnerable to these ailments during the sultry days of late summer and the subsequent rainy season.[36] William Hillary, who practiced on Barbados a decade later, followed Hippocrates' recommended procedure of studying the topography of the island as a key to its ailments. He measured the qualities of its atmosphere with a Fahrenheit thermometer, supplied from Amsterdam, and a portable barometer he had brought from London. He accounted for the physiological effects of heat and humidity in language similar to Bisset's. Heat was seen as expanding the fluids and relaxing the fibers of the body, thus expelling noxious materials from the skin. This could be observed in the small red lumps on the limbs, which people mistakenly believed were caused by the bites of mosquitoes. They were actually, Hillary claimed, signs of a healthy process of perspiration. Problems arose only if sweating was hindered by high humidity or inappropriate clothing. Hillary recommended that settlers abandon European styles of clothing, such as thick coats and waistcoats, and instead adopt the "banjan," a loose gown like those worn in Asia.[37] The banjan, or banyan, did in fact become quite fashionable in North America and Europe; its use could be rationalized on the medical grounds that it permitted free and healthy perspiration.[38]

The British colonies on the American mainland were closely connected with those of the Caribbean. Ships frequently traded between them, as well as linking them with the homeland on the other side of the Atlantic. It is not surprising that doctors in the North American colonies shared the general outlook of those practicing in the West Indies. Among them were two Scottish physicians working in partnership in Charleston, South Carolina: John Lining and Lionel Chalmers. Lining, originally from Lanarkshire, settled in Charleston in 1730.[39] He approached the question of the physiological effects of the climate by personal experimentation in the tradition of the seventeenth-century Paduan physician Santorio Santorio. Lining compiled meticulous records of his own intake of food and drink and his output in perspiration, urine, and feces from March 1740 through February of the following year. He weighed everything he consumed and all his evacuations; he weighed himself twice every day (on rising and before going to bed); and he also recorded his pulse rate. He combined this with a detailed record of the weather, using instruments to measure the barometric pressure, temperature, rainfall, and atmospheric humidity. Lining's journal was published in the *Philosophical Transactions* of the Royal Society. The author expressed the hope that his record would illuminate "the Changes produced in our Constitutions, disposing us to such and such Diseases, in certain Periods of the Year."[40]

While Lining approached the topic through narrowly focused—not to say obsessive—experimentation, his colleague Chalmers developed its wider social dimension. In his *Account of the Weather and Diseases of South Carolina* (1776), Chalmers explained that people were essentially the same everywhere, and "not otherwise to be distinguished from each other, than so far as they may be of more firm or feeble habits, according to their various climates."[41] This made the issue of climate fundamental to the health, welfare, and prosperity of the American population. Chalmers cast an anxious eye over the apparently unhealthy aspects of the Charleston milieu: the marshes with their mephitic stagnant water, the unwholesome fogs and dews, and the seasonal hazards of heat waves, tornadoes, and hurricanes. He shared the common belief among the colonists that the air would be improved by clearing forests and bringing more land under cultivation. So long as forests continued to surround his town, he wrote, the stagnant air "in those close recesses . . . renders them more proper for the habitations of wild beasts than of men." But not all social development was welcome: increased luxury and dissipation would weaken people's resistance to disease. Chalmers was particularly concerned about tea and coffee drinking, which, he fretted, "cannot fail in having ill consequences, in some constitutions, particularly during the relaxing heat of summer."[42] British writers such as George Cheyne had already developed the theme of the bad consequences of luxury and fashion for personal health. Tea and coffee were frequent targets for censure; these beverages—along with wearing fashionable clothing, dancing, and congregating in crowded rooms—were thought to increase people's susceptibility to airborne diseases. Chalmers was echoing this moralizing tendency in the British discourse of public health and reorienting it to the climatic situation of the colonies, where the dangers of lax behavior were heightened by an unfriendly environment.

The castigation of extravagance and immoderation, especially in female behavior, became a standard topic of medical writings about the hazards of colonial climates. As has been mentioned, Hillary criticized women for disregarding his advice not to exercise too vigorously. Their fondness for dancing was putting their health at risk, he warned—a point echoed by James Johnson in India in the early nineteenth century.[43] Even after the American colonies gained their independence, male observers (both American and foreign) continued to criticize women for indulging in habits that increased their vulnerability to the diseases of the climate. William Currie, a Philadelphia physician writing in the early 1790s, declared that women's illnesses were due to their drinking too much tea, breathing the air of confined spaces, frequently changing their dress, "and the alternate vicissitudes from heat to cold, to which fashion, and the love of pleasure, expose them." Currie broadened his reprimand to include young women who read "Love-inspiring Novels," who "not only impair their constitutions, but pervert their imaginations, and corrupt their morals to such a degree, that they are ever after rendered unfit for the offices of domestic life."[44] In the following decade, the French visitor Constantin François de Volney cast a sardonic European eye over the unhealthy indulgences of the Americans. All of them, he said in 1804, "live in a state of perpetual indigestion extremely favourable to catching colds." But he criticized women especially, whose susceptibility was said to be increased by light, fashionable clothing, "overheated apartments, balls, tea-parties, and featherbeds."[45]

These condemnations by male observers of women's behavior suggest that a general moral anxiety was sharpened by what were thought to be the climatic hazards of the colonial situation. In Britain, men resented women who were acting independently and enjoying new recreations, and doctors took it upon themselves to tell women that they did so at risk to their health. These dangers were thought to be increased in a setting in which the passions were likely to be less restrained than in the temperate homeland. Writers on climate and character generally agreed that hot weather lessened the inhibitions on sensuality and the other passions. In this respect, "relaxation" was both a physiological and a moral problem. The same circumstances that would loosen the bodily fibers and expand the fluids would also reduce conscious restraints on feelings and behavior. It was therefore thought particularly important to uphold rigid moral standards. A number of writers on tropical medicine emphasized this imperative.[46]

There was also, however, an underlying assumption that some sort of adaptation to local conditions was necessary. Medical writers did not advocate abandoning moral constraints, but they did often endorse the adoption of at least certain local habits. Thus, Sloane and Hillary advised settlers to relinquish their habitual clothing and assume a garb more suited to the climate. They urged the colonists not to exercise as vigorously as they were accustomed to at home. The rationale was to allow the individual to become acclimatized. It was assumed that a settler from Europe would gradually adjust to a tropical climate, if nature was allowed to do its work. Just as plants and animals were thought to be transplantable to distant places, so people were expected to be modified by the forces of nature itself to fit the climate to which they relocated. This was sometimes called "seasoning."[47] The great naturalists of the Enlightenment, including Carolus Linnaeus and Georges Louis Leclerc, Comte de Buffon, experimented with relocating plants and animals from the tropics to European institutions. They believed that climate would be the means by which natural forces would fit the organisms to their new circumstances.[48] Human beings were thought capable of a similar adjustment, even to the extent of changing their skin color, provided they allowed nature to act on their bodies. Everyone knew that Europeans became darker after they lived for a while in the tropics, and darker-skinned people were said to have become paler when relocated to Europe. It was widely reported that Portuguese settlers in West Africa had darkened over several generations, to the extent that they were now indistinguishable from natives. Writers on the phenomenon presumed that climate was the agent of these changes.[49]

The "seasoning" perspective generally held sway in tropical medicine, until it began to be challenged by notions of racial immutability just before the turn of the nineteenth century. Medical writers agreed that new arrivals in the tropics were the most vulnerable to the local diseases. If they survived a year or so, their chances thereafter would be much improved by having adapted to the conditions. James Lind, who as a naval surgeon pioneered methods for preventing scurvy on British ships, wrote his *Essay on Diseases Incidental to Europeans in Hot Climates* (1768) as a manual for settlers and soldiers in the tropics. He stressed that the climate outside Europe had frequently proved fatal to colonists, and that even adoption of a temperate mode of life provided no guarantee of survival. Every country, however, had its healthy places and its relatively healthy seasons. The best advice was for settlers to evacuate during the hazardous months to locations with more healthy air, at least in the first year or until they had become acclimatized.[50] An army physician, John Hunter, drew upon the work of doctors in Africa and India to address the specific situation of soldiers in the West Indies in his *Observations on the Diseases of the Army in Jamaica* (1788). He painted a shocking picture of army losses to disease during recent military campaigns, estimating that up to one-third of the members of active units were unavailable for duty at certain times because of sickness. In the course of a year, approximately one-quarter of the troops in service in Jamaica died of disease. These losses could barely be replenished by new recruits from Britain, especially because the new arrivals were particularly likely to succumb. Advice to soldiers to avoid intemperance could do little to meliorate this dire situation. The only remedy was to allow newly arrived troops to acclimatize gradually in the most healthy places that could be found. Duly seasoned, they would have at least a fighting chance of resisting the onslaught of the diseases that felled so many of their comrades.[51]

The expectation that Europeans would adapt to tropical climates did not mean they were supposed to be entirely passive in their occupation of new settlements. Although seasoning was supposed to occur by allowing the forces of nature to work upon the bodies of settlers, there was also scope for active intervention to alter the environment. Improvement of the air was seen as a way to help nature exert its beneficial effects, as Sir John Pringle influentially urged. His *Observations on the Diseases of the*

Army (1752) recommended that military camps be sited away from marshy ground, to avoid the "putrid miasma" it emitted. As we have already seen, Pringle energetically advocated a whole range of environmental interventions in Britain to improve the atmosphere surrounding human habitations. Similar improvements were also advocated in overseas possessions, where draining of swamps and clearing of forests were urgently demanded to improve the healthiness of the atmosphere. Lind noted that stagnant water and marshes, even in England, produced vapors that were noxious to health; they were necessarily much more hazardous in the tropics. Equatorial Africa was notorious for its hot and swampy air, but Lind was confident that "if any tract of land in Guinea was as well improved as the island of Barbadoes, and as perfectly freed from trees, shrubs, marshes, &c. the air would be rendered equally healthful there, as in that pleasant West Indian island."[52] The Portuguese had already shown, he claimed, that a settlement on the Congo River could be as healthy as anywhere, once its surrounding trees were cleared.

Settlers and observers in many British overseas possessions shared the belief that clearing forests and marshes would improve the quality of the local air. In the homeland, the fact that the climate was seen as a gift of providence did not mean that it could not be improved; in many colonial settlements, it was thought imperative that it should be. Chalmers looked forward to the time when improvements in the vicinity of Charleston would allow refreshing breezes from the ocean to circulate more easily. Hunter wrote that "noxious exhalations from wet, low, and marshy grounds" had been shown unhealthy "by repeated experience and observation in all parts of the world."[53] In view of this, army camps should preferably be located on hilltops or coasts, and it should be a military priority to clear and drain the land near existing sites. Of course, the heavy work would usually not be done by British settlers or soldiers themselves. Both Hunter and Lind made it clear that African slaves would be used to clear land in Africa and the West Indies.[54] So while Europeans prided themselves on their capacity to improve on nature, which they believed placed them at the apex of human civilization, they often made use of the labor of slaves to get the job done.

By the early nineteenth century, doubts began to be voiced about the idea that settlers would become seasoned to the climate in which they lived.[55] James Johnson, in his *Influence of Tropical Climates* (1813), insisted that the superiority of human beings over animals lay in the ingenuity of their minds, not the pliability of their bodies. He denied that humans shared animals' natural ability to adapt to their environment. Skin color would not in fact change, even over several generations. To say that it would seemed, to Johnson, to give comfort to the "gloomy doctrine" of materialism. He insisted that it was particularly unwise for Europeans to try to acclimatize by mimicking native customs, which "in reality, have ignorance, superstition, or even vice for their foundation."[56] They must trust to their intellectual capacity to fit them for life in the tropics, not hope that their bodies would adjust naturally. The colonists' strongest ally was their moral and mental superiority over the natives, not a natural ability to become seasoned to the prevailing conditions. Johnson told his readers that the inherent racial superiority of Europeans was the key to their ability to settle anywhere in the world. His outlook was consistent with the growing conviction among British intellectuals at the time that racial characteristics embraced mental as well as physical qualities, and that they were an immutable inheritance of the different strains of humanity.

Even while the notion of inherent racial differences took hold, however, European settlers did not completely abandon the hope that they could eventually adapt to life in tropical conditions. In the early nineteenth century, settlers in the American West held to the faith that they would become acclimatized, even sometimes worrying that the process would compromise their racial identity as white people.[57] In the eighteenth century, the prevailing assumption was that, given time, nature would fit people for the climate in which they lived. Underlying this belief was a sense that human nature linked people to their physical environment through bodily experience. As creatures of flesh and blood, human beings were inevitably affected by temperature and other atmospheric qualities, which penetrated their bodies and altered the rigidity of fibers and the velocity of fluids. These changes were thought to lie behind the alterations in people's health and passions that had repeatedly been catalogued in studies of weather and climate in many parts of the world. Whatever state of social development they enjoyed, human beings would inevitably remain subject to nature. This being so, climate would be the means by which nature would exert its unavoidable influence over European settlers in the tropics. This did not mean that the colonists should be entirely passive in their new environment. There was no escaping the effects of nature, but nature could be molded to exert its effects in a more desirable way. Hence the programs for improving the quality of the environment in the colonies by reshaping the landscape around settlements. Taking control of their physical surroundings by draining swamps and clearing forests allowed European settlers to enroll nature as an ally in their campaign to civilize the world around them and ease the process of acclimatization.

America: Climate and Destiny

The debate about climate and civilization engaged philosophers, historians, medical writers, and settlers themselves. It was conducted across a wide geographical range, frequently making comparisons between the British homeland and its far-flung colonial outposts. Climate was invoked as a way of trying to account for the diversity of humanity and for the many stages people seemed to occupy on the ladder of social development. It was thought to affect the physical dimension of human nature, requiring the adjustment of manners and laws to address its undesirable consequences. The British colonies in North America faced the conditions of their own climate with an outlook shaped by this debate. From the time when they first encountered it, the American continent had presented a challenge to Europeans' climatic expectations. In the eighteenth century, it also provided a focus for the debate on human history and the environment. Before and after the United States gained its independence, writers in Europe and America discussed how physical circumstances would shape the destiny of this society. Enlightened thinking about the relations between climatic conditions and the progress of civilization was of obvious importance for those seeking answers to the question. European writers were often read as denigrating American nature by suggesting that its climate had stunted the growth of animals and native human beings and by implying that it would limit the degree to which civilized society could develop there. American writers spoke up for their natural environment—and to some extent for their native peoples—against these strictures. They acknowledged differences between European and American climates, while minimizing the disadvantages and maximizing the advantages of the latter. Particularly after independence, Americans defended their natural environment as a support for the building of the nation.

In this connection, Europeans and Americans gave particular prominence to the idea that the climate was being changed by the consequences of colonial settlement. The clearing of forests and the cultivation of land by agriculture were almost universally said to have had measurable effects on the climate since Europeans first landed on the continent. These effects were also said to be noticeable in other colonial outposts, such as tropical islands; but they were emphasized with unparalleled regularity by commentators on America. There were two main reasons for this. First, writers on both sides of the Atlantic wanted to believe that nature was being civilized in the New World. America was a great project, in which many Enlightenment hopes were invested, and it was expected that the taming of its wilderness and climate would follow from the expansion of its settlements. A second reason for the expression of these hopes in relation to America was the fate of its native peoples, who had suffered a disastrous decline in population since the arrival of Europeans, primarily due to epidemics of such fatal diseases as smallpox. The continuing decline of the natives gave particular urgency to consideration of the role of climate by the settler population. It was hard to resist the assertions of writers in Europe who claimed that the American climate had had a damaging or weakening effect on the natives. Though American writers often denied that this was so, and did what they could to defend the natives' reputation for vigor and strength, they also sought reassurance in the belief that the climate was changing. It was important to assert that whatever its undesirable consequences in the past, its effects were no longer to be feared. These two factors inclined American writers to see the climate as having been significantly transformed since European settlement began. The climate Americans claimed as a national asset was one they believed they had molded—and were continuing to mold—to meet the requirements of their civilization.

The weather in the New World had posed a conundrum to settlers from the beginning. Europeans venturing across the Atlantic quickly noticed that American locations were much colder in winter than the corresponding latitudes in Europe; in summer they could be hotter and considerably more humid. To determine the prospects for settlement, it was essential to find out what local conditions were like and how they varied with the seasons.[58] By the late seventeenth century, British methods of systematic weather recording were being used to chart conditions in America. The Royal Society welcomed reports from the other side of the Atlantic that used these methods. In the 1690s, letters by John Clayton about the natural history and climate of Virginia were published in the *Philosophical Transactions*. In a Hippocratic vein, Clayton, who had been a minister at Jamestown in the 1680s, noted how sudden changes in the Virginian weather affected the health of the inhabitants. Thomas Robie kept a weather journal (though without instruments) from 1715 to 1722 at Harvard College in Massachusetts, where he served as tutor, sending it later to William Derham to share with the London virtuosi. James Jurin's invitation to meteorological record keepers in the early 1720s met with a response from Isaac Greenwood, a professor at Harvard, whose proposals for compiling a "natural history of meteors" appeared in the *Philosophical Transactions*. A few years later, Paul Dudley, a judge of the Massachusetts Superior Court, sent another weather journal to the Royal Society covering the years 1729 to 1733.[59]

FIGURE 18 · "A View of the Waterspout Seen at the Entrance of Cape
Fear River." An American weather wonder shown in the frontispiece
to Thomas Branagan, *The Pleasures of Contemplation* (Philadelphia, 1817).
Courtesy American Antiquarian Society, Worcester, Massachusetts.

These colonial observers looked to London for accreditation and publication of their research. They participated in a transatlantic trade, in which
specimens and written descriptions were centralized in the metropolis and
paid for in the currency of social prestige minted by such institutions as the
Royal Society. Naturalists and weather observers in the American colonies
attached themselves to the far-flung networks by which this knowledge
was accumulated in the imperial capital. They also imported meteorological instruments from the metropolis. After a series of mishaps attending
their transportation across the Atlantic, barometers and thermometers
began to be made available in the colonies by the 1720s. They were soon
used to report atmospheric measurements to London. Harvard received a
portable barometer for its instrument collection in 1727, and it was used by
Greenwood for observations. John Winthrop, Greenwood's successor in the
Hollis chair of natural philosophy at Harvard, compiled a record of temperature and pressure in Massachusetts from 1743 to 1747, which he sent to the
Royal Society.[60] At around the same time, Lining was using instruments in
South Carolina. The following decade saw measuring apparatus being used
systematically by Chalmers at Charleston and by Hillary on Barbados. By
the beginning of the nineteenth century, Volney noted acutely that Brit-

ish and American meteorologists shared the same quantitative approach to their subject when, "conformably to the national genius, [they] reduce every thing to direct and systematic calculations." [61]

American weather observers shared methods with their British colleagues and echoed many of their preoccupations. Their comments often took it for granted that the British climate was the norm and that American conditions were hazardous insofar as they deviated from it. Thus, extreme temperatures were of great interest, particularly when they could be quantified. The *Philosophical Transactions* published accounts of winters at Hudson's Bay when it was so cold as to freeze the mercury in the thermometer tube, and of summers in Georgia when 102 degrees was recorded on the Fahrenheit scale. In the summer of 1752, Chalmers measured the heat in his kitchen in South Carolina at 115 degrees. He anticipated that his record of the occasion might "not displease the curious," as no register of such a hot season had previously been published. [62] Such extreme departures from the temperatures usual in Britain gave rise to serious worries about their effects on health. Chalmers noted that the hot and humid summers in Charleston were particularly dangerous times for fevers. [63] Similarly, Lind commented that the hot locations in North America were the ones where settlers' health was particularly precarious. Like other medical writers, these two often made comparisons between the American colonies and British settlements in the tropics. But there were also writers, such as William Robertson, who emphasized the prevalence of cold in America and who traced its negative effects on the health and vigor of the inhabitants. Whether perceived as too hot or too cold, it was the American climate's differences from the British climate that aroused anxieties about sickness.

A further aspect of the exoticism of the American climate was its apparent fertility in atmospheric wonders. Reports of tornadoes, waterspouts, hurricanes, thunderstorms, and other prodigious meteors appeared frequently in metropolitan and colonial publications. In colonies where Puritanism was influential, such phenomena continued to be regarded as divine portents well into the eighteenth century. The arguments about their reducibility to natural law — arguments that, we have seen, swirled around the storm of 1703 in Britain — resurfaced periodically. [64] In Boston, John Winthrop championed the naturalistic view, initially in connection with an earthquake that struck New England in 1755, then on the occasion of a comet in 1758, and later in descriptions of a series of fiery American meteors sent to the Royal Society in the 1760s. [65] The problem was that the more the descriptions of such anomalies were elaborated, the more difficult it was to assimilate them to the regular order of nature. Benjamin

FIGURE 19 · Benjamin Franklin and lightning.
The American philosopher with accoutrements of
his electrical experiments, from a mezzotint
portrait of the 1740s. Courtesy of Science and
Society Picture Library, London.

Franklin's work on lightning, beginning in the 1740s, aimed to reduce the phenomenon to natural law and lessen its dangers with the use of lightning rods; but it also made everyone aware of the violent electricity of the American atmosphere. As the fame of Franklin's accomplishments spread, it came to be generally accepted that the air was more electrically charged in America than in Europe.[66] These wonders added to anxieties about the hazards of the American climate and to the urgency of the task of taming it.

In the context of such worries, the idea of a transformation of the American climate assumed considerable prominence. It was widely asserted that settlement and cultivation of the American landscape by Europeans was bringing the weather into line with the temperate ideal of the Old World. The cutting of American forests, especially, was said to be moderating seasonal extremes of temperature.[67] Comparisons were made with the changes said to have occurred in the European climate since ancient or medieval times. Centuries of civilization in Europe were thought to have ameliorated the brutal weather conditions recorded by classical writers. European settlers in North America believed they were effecting a similar transformation in a much shorter period. They thus ignored the role of the native peoples in shaping the landscape of the Americas, dismissing their agriculture as insignificant and minimizing the scale of their settlements. It was said that the wilderness had remained unimproved until the arrival of Europeans, who had begun the process of civilizing the American environment and softening its climatic extremes.

In the second half of the eighteenth century, this idea came to be widely credited by European and American writers. Pehr Kalm, a Swedish follower of Linnaeus, raised the matter in his account of a visit to the New World in 1749. He was told by old people he encountered in New Jersey and Quebec that the winters had previously been longer and harsher, and the yield of the wheat crop much less. He was also told that people ascribed the climatic amelioration to the clearing of trees, which allowed the sun to act more directly on the soil.[68] In a paper read to the American Philosophical Society in Philadelphia in 1770, Hugh Williamson, a Philadelphia physician, claimed that eastern North America had become significantly more temperate after settlement, especially because of deforestation. He stated that removing forests would lessen the cold of winters and the heat of summers, with consequent benefits for people's health.[69] The assertion was picked up by European writers such as Buffon, who included the draining of marshes among the causes of the alteration. In his *History of America*, Robertson contrasted the industrious cultivation of the American landscape by European settlers with its supposed neglect by the native peoples and pointed out that "when any region lies neglected and destitute of cultivation, . . . the malignity of the distempers natural to the climate increases."[70] Robertson's fellow Scot James Dunbar agreed that "by opening the soil, by clearing the forests, by cutting out passages for the stagnant waters, the new hemisphere becomes auspicious, like the old, for the growth and population of mankind."[71]

In America itself, the belief that the climate was being changed was taken to heart by the settler population as the United States claimed its independence. Thomas Jefferson, in his *Notes on the State of Virginia* (1787), reported as a result of his own research among the settler population that "both heats and colds are become much more moderate within the memory even of the middle aged." Late in his life, he called for a national network of weather observers to compile prolonged observations in order "to show the effect of clearing and culture towards changes of climate."[72] In

Philadelphia, William Currie pointed to the extent of the native forests to explain why the weather was originally found "less agreeable" in America than at corresponding latitudes in Europe. He compared the situation to that in the ancient civilization of China, where, by extensive alterations of the landscape, "the air, in very unfavourable situations, has been rendered exceedingly wholesome." [73] Italy, Germany, and England had improved their climates within historic times, according to Currie. The same could be expected in the United States: "When in the course of time, this continent becomes populated, cleared, cultivated, improved, and the moisture of the soil exhausted, . . . the bleak winds will become more mild, and the Winters less cold." [74]

By the last decade of the century, a widespread consensus had formed that deforestation and other improvements had already reduced the severity of winter cold in America and moderated other climatic extremes. The topic seems to have attracted experimental investigation after having surfaced in popular consciousness. Learned authors debated its magnitude and questioned whether it was altogether a good thing. Benjamin Rush, the most famous Philadelphia physician of the era and professor of chemistry at the University of Pennsylvania, addressed the question in a paper published in 1789. He accepted that "accounts which have been handed down to us by our ancestors" gave reason to believe that the climate had changed.[75] But the question was tricky to specify empirically because of the paucity of exact records from the early stages of colonization. Rush suspected that memories of the elderly, the source of Kalm's information, were unreliable, perhaps because people's perceptions of heat and cold altered as they aged. He concluded that there was no decisive evidence that winters had been colder before 1740 than after, but he agreed that the seasons had tended to merge into one another and the weather had become more variable in recent years. He accepted Williamson's assertion that clearing of forests and cultivation of the land were largely responsible for this. A few years later, the historian and physician David Ramsay made another assault on the question in his *Sketch of the Soil, Climate, Weather, and Diseases of South Carolina* (1796). Ramsay drew upon the long tradition of weather observations in Charleston, comparing contemporary records with those compiled by Chalmers in the middle of the century. He concluded that both maximum and minimum temperatures had moderated over the period, but that it was too soon to say whether this represented a long-term trend. He nonetheless reasserted the basic assumption that improvements in the natural environment would bring permanent benefits in terms of climate and health: "The advantages resulting to the temperature of the air, and to the healthiness, as well as to the appearance of any country, from the art of man, inhabiting and cultivating it, are inconceivably great. We may, therefore, indulge the hope that our [climate] is progressively meliorating from permanent and encreasing causes." [76]

Samuel Williams's *Natural and Civil History of Vermont* (1794) reported that climatic transformation was "so rapid and constant, that it is the subject of common observation and experience. It has been observed in every part of the United States." [77] Throughout the country, winters had become shorter, summers less intensely hot, and the weather in general subject to more rapid variations. Williams, a member of the American Philosophical Society in Philadelphia and the Palatine Meteorological Society in Germany, also set out to explore the issue experimentally. He measured soil temperatures in uncut woods and in open fields and concluded that deforestation had measurably warmed the soil. He also measured the rate of evaporation of water from leaves, trying to estimate how much atmospheric humidity was reduced by removing forests. He understood that discoveries by Joseph Priestley and others had shown the importance of vegetation in restoring the air's suitability for respiration. His final conclusion was finely balanced, reflecting the consensus that change had happened but entering some reservations about its overall benefits.[78] Jeremy Belknap, a Congregationalist minister whose *History of New-Hampshire* (1812) investigated the question a couple of decades after Williams, agreed that trees had the virtue of purifying the air. He argued that New Hampshire owed the good health of its population to its rugged environment, including its forests.[79] Like Williams, Belknap was aware that trees had been shown to contribute to the healthiness of the air. These writers pointed the way to the emergence of a custodial attitude to the American forests, following in the wake of a conservation movement already established in the Caribbean.[80] But the campaign to preserve the forests emerged only slowly in the United States. Overall, the conviction prevailed that cutting down trees and cultivating the landscape changed the climate for the better. When Volney visited in the late 1790s, he reported that these changes were recognized by everyone and "have been represented to me not as gradual and progressive, but as rapid and almost sudden, in proportion to the extent to which the land is cleared." [81] Americans believed that they were civilizing their own nature, and doing so more rapidly than the Europeans had. In the nineteenth century, as agriculture spread to the Midwest, clearing the forest was regarded as the first step to rendering the land healthy and productive. On the Great Plains, it was commonly said that "rain follows the plough"—that cultivation would change the climate in a way that would favor further settlement. This notion had first become established in the original thirteen colonies in the period when the nation was born.[82]

The idea of climatic transformation appealed to Americans' sense of their national destiny, their faith that providence was guiding the conversion of the wilderness into a civilized country. It also provided a way of responding to hostile remarks by European writers about the American environment and its effects. Buffon had led the campaign on this front with his assertion that nature in the New World had caused the degeneration of animal species transplanted from Europe. The same cold and humid conditions had stunted the growth and vigor of New World peoples, according to Buffon, who nonetheless saw a sign of hope in the prospect that the American climate might be changed to match that of Europe.[83] Robertson's strictures on the New World climate, particularly its effects on the natives, were especially severe. American nature had checked the growth of "the more noble animals" while encouraging odious reptiles and insects, "the offspring of heat, moisture, and corruption." The native peoples had been thwarted in their development, both physical and cultural, by the same cause. Afflicted with bodily weakness and a disposition to melancholy, they lacked the force of mind to take charge of their environment by domesticating animals or plowing the soil. Robertson did allow, however, that European settlers were taming the wilderness, which would permit them to escape the debilitating effects of the original climate and even offered the prospect of improving the character of the natives.[84]

Although Buffon and Robertson exempted European settlers from having had their constitution negatively affected by the climate, their criticisms were read as attacks on American nature, with worrying implications for the fate of the continent's civilization. Several American authors responded. In his *Notes on the State of Virginia*, Jefferson defended the climate of his state and insisted that it was already showing signs of being tamed by civilization.[85] In addition, he resisted the assertion that American climatic conditions had yielded degenerate forms of animal life. Jefferson had confronted Buffon personally on this matter in Paris, trying to clinch his point by sending across the Atlantic a specimen of an American moose, a mammal undeniably larger than any European equivalent. Nor would he accept that America had produced native human beings who lacked vitality and social sentiment; he insisted that Native Americans were not deficient by nature—they were simply at an early stage of social development.[86] As recent scholars have noted, Jefferson's sympathy for native people was not extended to the Africans enslaved on his own estate and throughout the South. He adopted Hume's position that Africans were intellectually inferior to Europeans "by nature," and hence were likely to remain in the condition of slaves.[87] Apparently, it was more important to defend the reputation of Native Americans, presumably partly because they were products of the continent and its climate with which the new nation had cast its lot. As Charles A. Miller has put it, "Jefferson identified the human nature of America with its natural history, thus establishing a bond with the Indians that was inconceivable with the Africans."[88] A similar defense was mounted by Samuel Williams in his *History of Vermont*. He criticized Buffon as an armchair philosopher ignorant of the realities of American natives. "No such animal was ever seen in America, as the Indian M. de Buffon described in Paris," Williams insisted. It was not true that native people lacked sensibility, energy, or sexual drive: "Nature is the same in the Indian, as it is in the European."[89] Native Americans might have lacked cultivated morals, but they possessed the basic virtues of love of country and fierce independence — virtues that Williams thought all Americans should be proud to embrace. In contrast to the corrupting effects of European luxury and indulgence, the simple virtues of New World natives were said to derive from their closeness to American nature. These were therefore virtues that all inhabitants of the new republic might hope to share.

Jefferson, Williams, and others defended the native peoples because they identified them with the American environment.[90] The climate that had shaped them would inevitably set the conditions for the new nation's future, modified though it would be by the accumulated effects of cultivation. Volney, who fled to the United States in 1795 as a refugee from the revolutionary regime in France and left disillusioned three years later, wrote that the Americans had a peculiar pride in their climate. As far as he could see, it was altogether less desirable than that of Mediterranean countries, but Americans stubbornly defended its qualities. Volney thought that this could only be ascribed to self-interest and the simple fact that people got habituated to the conditions in which they lived. Their judgment was distorted by imbibing "a physical and moral atmosphere, which we breathe without perceiving it."[91] Americans themselves spoke of their newfound liberty as a "climate" or an "atmosphere," or as the direct result of the workings of nature on their moral constitution. Williams wrote that while European monarchs delayed the progress of reform, "nature was establishing a system of freedom in America."[92] Currie compared the ideal climate that Americans could expect to enjoy when the land was cleared with the political liberty they were already experiencing, having "already, in a great measure, regained the native dignity of our species."[93] Americans were beginning to pride themselves on their climate, which they saw as the source of their virtues, especially the vigorous defense of liberty that had won their nation's independence.

Although it was articulated in arguments against European philosophers and developed in the context of the war of independence from Britain, this attitude was rooted in enlightened ways of thinking that had also found expression in British culture with the notion that climate could underpin national identity.[94] Americans understood their climate as integral to their destiny because they saw it as one of the means by which nature exerted its pervasive influence on human life at all stages of historical development. They were sometimes inclined to defend the virtues of the natives as products of the rugged American climate. Nature was thought to have given birth to heroically independent people who could inspire patriotic Americans as models of liberty. On the other hand, anxieties about the apparent weaknesses of the natives—which were unavoidable in view of their catastrophic susceptibility to diseases—could be allayed by the conviction that the climate was being changed by cultivation. European settlers prided themselves on having moderated the extremes of their weather by altering the landscape. Unlike the native peoples, they believed, they had taken hold of the environment around them and reshaped it to their needs. This would prevent them from sharing the natives' appalling fate.

Underlying these views about the American climate were attitudes to the relationship between nature and civilization that we have seen expressed in other contexts. The philosophers and historians who debated the role of physical and moral causes in history believed that nature was a shaping presence even in highly civilized society. But they did not believe that human beings were passive objects of natural forces. Rather, active intervention in the natural environment was an aspect of human nature itself. People—by their nature—acted on their physical surroundings. The same principle was at work in the commentaries of medical writers who advocated reshaping the landscape around settlements in the tropics. Civilized people could take action to redirect the forces of nature, for example by removing the sources of unhealthy air in marshes and forests. In doing this, they were not acting against nature, which enlightened thinkers would have considered impossible, but giving expression to an element in the natural constitution of humanity itself. Americans who asserted that they had derived their love of freedom from the environment of the New World were making a similar claim about the roots of human nature. Human beings were seen as the products of nature, even as they insisted on their prerogative to remold the surrounding milieu.

The Science of Weather

All these things which do not harmonize with one another do suit well with that lower part of creation which we call the earth, which has its cloudy and windy sky in some way apt to it.

ST. AUGUSTINE · *The Confessions*

You can't *build* clouds. And that's why the future you *dream* of never comes true.

LUDWIG WITTGENSTEIN · *Culture and Value*

AS WAS NOTED AT THE OUTSET, this is not a book about the history of meteorology as usually conceived. I have not been telling the story of how studies of the atmosphere assumed a scientific cast. There are already several books that give accounts of the development of meteorology in this period.[1] I have drawn upon them and refer to some of the same primary materials, but my basic aim has been different. I set out to explore how attitudes to weather and climate reflected experiences of the Enlightenment in Britain and its colonies. Pursuing the relations between these attitudes and cultural change, I have been concerned with general patterns of belief, not with systematic bodies of knowledge. I have not, until this point, raised the question of the scientific status of the ideas in question.

The question is, however, worth asking. It is worth considering whether a scientific study of the weather existed in this period and, if it did, what kind of knowledge it produced. I cannot answer the question comprehensively at the moment—that would require another book—but I think the evidence I have gathered suggests the outlines of an answer. In the previous chapters, I have noted how people began to think that the weather reflected the regular course of nature, and how they hoped at the same time that diligent observation would reveal its laws. I have discussed how British investigators forged a sense of the national climate, in which the normal pattern of weather was viewed as a providential gift to the nation's prosperity and health. By way of a conclusion, I shall reiterate these points and introduce some new information to show how those studying the weather conceived of the goals of their investigation and the extent to which they believed they had achieved them. As we shall see, the picture shows both successes and failures. Extensive records were certainly accumulated, but I have already noted a number of ways in which studies of the weather failed to attain their most ambitious aims. Indeed, it could be said that this has continued to be the case ever since the eighteenth century. Meteorology continues to this day to grapple with inherent uncertainties; because it often falls short of its predictive ambitions, its scientific credentials are still called into question.

We can begin by asking, what did people in the eighteenth century believe would constitute a "science" of the weather? The word *science* was used in this period to mean a systematic body of warranted knowledge, and was not necessarily confined in its application to studies of the natural world. As the quotations from St. Augustine and Wittgenstein indicate, philosophers have long raised doubts about the very possibility of a "science" of the weather, in the strict sense of the term, holding that atmospheric phenomena occupy a domain of inherent uncertainty. In the late seventeenth century, it was often said that inquiries into the atmosphere should aspire to meet Francis Bacon's call for a "natural history" of the air. This presented an alternative to the classical Aristotelian conception of "meteorology," which had been concerned with the appearances known as "meteors." Aristotle's meteorology comprehended all phenomena that occurred in the realm below the orbit of the moon, including comets, shooting stars, and effluvia vented from beneath the earth, as well as things that would later be considered truly atmospheric. Many of these unusual or preternatural entities would also find a place in a Baconian natural history of the air, but they would be accompanied by more routine occurrences such as those we would recognize as "weather." Robert Boyle apparently felt he was making progress toward fulfilling the Baconian project in his posthumously published *General History of the Air* (1692). Boyle's work included discussion of the temperature, pressure, and humidity of the air, along with its motions, its chemical effects, and its contamination by terrestrial effluvia. As has been mentioned, John Locke, who edited the work for publication, incorporated a portion of his own weather diary, indicating that a Baconian natural history could take the form of a chronological record of weather phenomena.[2] A few years later, the Edgiock diarist insisted that recording the properties of the atmosphere was the route to "a vast & extensive science."[3] Compiling a weather journal served the needs of natural history, marking a preliminary step toward the drawing of philosophical conclusions and the formation of scientific knowledge.

An additional aim of those who studied the weather, already expressed in the seventeenth century, was to predict it. Robert Hooke's instructions for compiling a weather journal, published in 1667, declared the goal of finding "laws" that governed the atmosphere.[4] This hope was boosted by Newton's accomplishments in celestial mechanics. It was expected that changes in the air could be made as predictable as the motions of planets and comets by the discovery of underlying laws. This expectation reflected the belief that divine providence expressed itself uniformly in natural processes; it provided a further inducement to engage in meticulous record-keeping over the long term. But the ideal was still unrealized in the early nineteenth century. More than a hundred years after Newton, no such laws of the atmosphere had been discovered. Richard Kirwan wrote in 1794 that the scientific method was "as yet in its infancy" so long as this remained the case.[5] The failure was something of a scandal, as if enlightened investigators were reduced to the condition of primitive men gazing in clueless wonder at the heavens. Luke Howard declared in 1818 that weather observers were still performing for their science the office undertaken for astronomy by the Chaldean shepherds. Notwithstanding the efforts of generations of such observers, he lamented, "Meteorology . . . is yet far from having acquired the regular and consistent form of a science."[6] Howard began to suspect it would never do so, that "from the very nature of the causes concerned," meteorology could never attain the predictive certainty of astronomy.[7] His worries on this score were to be echoed by other commentators in the course of the nineteenth century. Certainly, meteorologists of that period were well aware that the predictive goals of the previous century had not been achieved, though some thought they still might be if investigators would renew their efforts.

The consistent failure of eighteenth-century meteorology to accomplish what it had set out to do should feature in the telling of its story. While the steady efforts of many observers deserve to be acknowledged, it must be recognized that they were never able to explain changes in the weather convincingly or to forecast them reliably. The issue has been discussed in relation to a later period. In her recent account of Victorian weather fore-casters, Katharine Anderson has suggested that the history of meteorology reveals particularly well the problems surrounding science's establishment of its social authority, precisely because *this* science was relatively unsuc-cessful.[8] Similarly, for the eighteenth century, I have emphasized those re-spects in which the field fell short of its greatest ambitions, because they seem symptomatic of enlightened science encountering the limits of its capabilities. Thus, the failure of systematic research to predict the weather left the field open to prognostication by the traditional techniques of "weather-wising." The weather did not in fact conform to regularity, so fears of its extremes and anomalies continued to be voiced. Even the suc-cesses of eighteenth-century meteorology often had consequences that ran counter to the ambitions of enlightened intellectuals. Medical research showed the importance of an atmospheric sensibility that was not subject to rational control. The newly invented scientific instruments often seemed to be regarded by their users with a kind of superstitious awe. The short-comings of meteorology as a science, which Anderson has uncovered in the nineteenth century, appear to have been rooted in the previous era.

The historian and philosopher of science Thomas S. Kuhn placed meteorology among the Baconian sciences of the early-modern period, applying his own criterion of a discipline that had not yet found a ruling paradigm and was therefore obliged to proceed by steadily accumulating observations.[9] There is no doubt that Bacon's inductive method provided a model for this field of inquiry, but to invoke it does not do the whole work of historical explanation, particularly when it comes to the collec-tive organization of the enterprise. Scientific knowledge of the atmosphere requires information from dispersed observers to be concentrated at some central point. Bacon had in fact specified a hierarchical social structure within which any scientific investigation was to be conducted. But in the event, although individual weather observers were plentiful through-out the eighteenth century, coordination of their efforts was sporadic, with rather striking gaps in continuity. The Baconian enterprise seems to have been launched repeatedly—by Hooke, Boyle, and others in the 1660s, for example, and then again by James Jurin in the 1720s—only to slump into inactivity. In the 1770s, Henry Cavendish reported on the state of the Royal Society's meteorological instruments and recommended revised protocols for using them.[10] A weather journal began to appear regularly in the *Philo-sophical Transactions*. The following decade, observational initiatives were launched in France, Germany, and elsewhere on the European continent. Although these have been hailed as unprecedented accomplishments, the projects lapsed again after a few years.[11] In 1801, Kirwan repeated the call for groups across Europe to communicate their observations on a regular basis. In 1823, John Frederic Daniell made severe criticisms of the observing practices in the Royal Society. It appears that the enterprise of organizing meteorological observers had to be kick-started repeatedly every couple of decades. Realizing this, we might be inclined to question the assumption that these initiatives should be identified with one another as part of a continuous history. If the deep story is one of the steady progress of observational knowledge, why was it so difficult to maintain momentum? Why were there so many years when nothing seems to have been happening?

To begin to answer these questions, we need to place each individual initiative in its appropriate context. It is far from clear that initiatives as far apart chronologically as the 1660s and the 1830s should be considered parts of the same project. Hooke's publication in the 1660s of a standard form to record weather observations was consistent with Bacon's vision of the process of induction, which required instances of phenomena to be assembled in tables and subjected to the gaze of a superior intellect who could draw axioms from them.[12] Hooke understood the need for observers to use standardized instruments, but he largely relied on the market among commercial suppliers to bring this about. He was looking for observers to submit diaries of their observations that would extend, if possible, over fairly long periods. He was not concerned specifically with the timing of observations, nor with where they were made; the only qualification required of observers was prolonged local knowledge. In retrospect, from the vantage point of later organizational projects, the vagueness of Hooke's stipulations would come to seem unpardonably lax. But Hooke understood that the Royal Society operated in a situation in which individuals' contributions were voluntary and rewarded only in the currency of peer approval. Weather observers were almost all drawn from the class of gentlemanly virtuosi; they were accustomed to setting their own hours for their avocations, not to being instructed as to when they should do their work. They were oriented toward the enterprise of chorography, or local descriptive geography, even while they also participated in the wider republic of letters.[13] Hooke appreciated that the project he was launching had to rely on this culture of local expertise and virtuoso sensibility if it was to achieve anything.

Turning to the other end of the period, it is clear that Daniell and other meteorological reformers of the early nineteenth century were striving for a much higher degree of precision in measurement and a regularity in the timing of observations that would bring work methods into line with the ideals of the industrial age. It seemed scandalous to Daniell that the variation in timing of the Royal Society's daily observations was "obviously regulated by nothing but the observer's night-cap." [14] Another reformer, Sir John Herschel, looked to men subject to military or naval discipline to find observers who would keep more regular hours. Instructing naval officers on the procedures to be used to compile meteorological registers, Herschel noted, "Irksome as it may be to landsmen to observe at 3 A.M., the habits of life on shipboard render it much less difficult to secure this hour in a trustworthy manner." [15] Maritime routines, military habits, even the regular devotions of members of religious orders, provided useful precedents for the more meticulous regularity that Daniell and Herschel sought to introduce into weather recording. And much else had changed in the century and a half since Hooke had published his invitation. Learned societies had developed a more systematic concern with the climate as a national asset having crucial effects on agricultural productivity and the health of the population. Much more accurate instruments were available, through improvement of the skills of instrument makers and a considerable expansion in the market. The culture of precision measurement had been placed at the service of nation-states in such enterprises as the cartographic surveys of the eighteenth century. Graphical methods had been devised by Alexander von Humboldt and others to map physical and biological variables across extended topographical domains. Herschel and others sought to capitalize on these developments to organize weather reporting on an unprecedented scale through the British Association for the Advancement of Science, beginning in the 1830s. They perceived the possibility of mobilizing a network of disciplined observers dispersed across an extensive geographical area. Meteorology, along with studies of terrestrial magnetism and the oceanic tides, became a means for the scientific mastery of geographical space. Men were to be employed to engage in the "extensive fagging" that such ambitious projects demanded. [16]

Situating these organizational initiatives in their specific historical contexts, we can begin to see what factors determined their successes and failures. Historians of science in recent years have repeatedly noted the social dimension of the creation of natural knowledge, how it relies on the distribution of skills and the coordination of actions as well as on the dispersal of material instruments. Histories of meteorology have already given significant attention to instruments, especially in relation to standardization of scales of measurement and advances in precision. [17] But there is still a need to widen the angle of view, so that more can be learned about how apparatus was actually used. Attention needs to be focused not just on events at the center of the observing network, but also on what was happening at the peripheries. This means considering both the ambitions of those who sought to recruit observers and the characteristics of the individuals they tried to enroll. Weather observers such as John Locke, Gilbert White, or even Luke Howard were largely following their own agendas, indulging their own gentlemanly curiosity in accordance with their own routines. They were, according to Howard, "gentlemen possessing . . . domestic habits, [aiming to] . . . agreeably fill up a portion of their daily leisure."[18] Their records, as Kirwan acknowledged, were inevitably subject to interruption "by death, sickness, or the common cares of life."[19] From the vantage point of Herschel and his colleagues in the age of scientific and political reform, eighteenth-century practices seemed to be mired in stagnation and amateurism. Historians should not succumb uncritically to this retrospective point of view, but some recognition of the limits of the organizational infrastructure of the time does seem appropriate.

Lacking continuous coordination, weather observers in the eighteenth century set their own objectives. As we have seen, their motivations were partly personal and partly connected with a sense of public duty. They shared the conviction that the way to make progress was through the compilation of a "longitudinal" (chronologically extended) record at a particular place. Their object, as Howard put it, was "a knowledge of the peculiar features of their own climate, and of the facts which, properly arranged, would form its *history*."[20] Even while they hoped for some form of coordination with other observers, their main focus was on maintaining the record at their own location in order to grasp the temporal connection between weather events. In 1794, Kirwan was still proclaiming that meteorology would mature as a science when the problem of the succession of atmospheric phenomena was solved, "the order in which they present themselves and succeed each other" having hitherto eluded research.[21] This chronologically extended kind of knowledge continued to draw the interest of investigators well beyond the eighteenth century. Howard's decades-long search to tie down the moon's influence on the weather exemplifies it, as does Kirwan's study of the patterns of the seasons. Kirwan devoted considerable effort to finding rules for forecasting a dry or wet summer or autumn from the weather of the preceding season. The project necessarily required a prolonged series of observations; he made his own at his home in Dublin in the 1790s, and he also drew on those of John Rutty and Thomas Barker from previous decades.[22]

At the same time, Kirwan showed an interest in the spatial dimension of weather phenomena, anticipating in some ways later developments in the science. He complained about observers who failed to pay attention to conditions at distant locations and called for a *"conspiracy,* if I may so call it, of all nations" to compile simultaneous observations from across Europe.[23] He appreciated that standardization of instruments was of vital importance to this campaign, which he saw as an appropriate expression of an advanced stage of enlightenment on the European continent. But he had little idea what form atmospheric phenomena would take if they were studied over a geographically extended area, or even what size they would turn out to be. Pulling together some of the available data on the subject in 1801, Kirwan mentioned the mapping of the Gulf Stream by Benjamin Franklin and others in the 1780s. Perhaps atmospheric currents would be found to have a size and speed similar to those in the oceans? Perhaps they could be traced in the record of barometric measurements made at the same time in different parts of the Russian empire in the 1760s? Or perhaps much larger-scale motions of the air could be found, at a truly global scale, revealed in the record of winds experienced a few days apart in St. Petersburg and in the Pacific Ocean? In view of the uncertainties of what Kirwan admitted in 1801 was "this obscure subject," he seems to have felt that more secure knowledge of the geographical aspects of weather could be sought by revising the ancient doctrine of climates.[24] Fourteen years earlier, his *Estimate of the Temperature of Different Latitudes* (1787) had studied how average temperatures at various places could be predicted from the angle of incidence of sunlight on the earth's surface. Kirwan still felt the lure of the classical concept of climates corresponding to zones of latitude. It was more straightforward to approach the problem of relating climate to geographical space in this manner than by grappling with the protean phenomena of weather.

Kirwan stood at a turning point in the development of geographical knowledge of the atmosphere. While looking back to ancient ideas, he also pointed forward to what happened later. In the course of the nineteenth century, large-scale movements of the air were mapped with considerable success. Storm systems, anticyclones, and (in the early twentieth century) weather fronts came to be identified, their motions charted and eventually predicted. These developments depended on the crucial technology of the telegraph, and later on wireless communication, which allowed for the rapid collection of simultaneous observations from widely dispersed points. It was only possible to begin to map such weather phenomena when human messages could move more rapidly than the air itself. A series of visual techniques was then developed to trace on paper the motions of air masses shortly after they had occurred, and eventually to offer predictions of how they would behave in the coming hours.[25] At the beginning of the nineteenth century, this was simply inconceivable. For all that they sought to collect observations from dispersed sites, meteorologists of that era had little idea of the spatial scale of the phenomena they were searching for. And as regards the dimension of time, their attention was focused on changes measured in days, months, or seasons rather than in hours. Only in the 1850s, with the establishment of telegraphic networks to convey weather information, did meteorologists begin to consider events happening on a much shorter timescale across a much larger expanse of geographical space.

The significance of techniques for mapping weather systems in nineteenth-century meteorology, and their prominence in the public profile of weather science today, can lead to an overemphasis on the eighteenth-century developments that pointed in this direction. For example, Franklin's work on the Gulf Stream was preceded by his tracking of the path of a nor'easter storm up the eastern seaboard of North America. The insight was an original one, recalled later as a pioneering attempt to understand the dynamics of storms; but it was not typical of studies of the weather in the period.[26] Insofar as people of the time were investigating the spatial dimension of weather phenomena, they were doing so without any inkling of the later theories of atmospheric systems. Their efforts generally require reference to contexts other than meteorology to be understood historically. For example, Edmond Halley's important map of the trade winds, produced in the 1680s, served the enterprise of improving techniques of oceanic navigation, a project that also issued in his simultaneous work to map magnetic variation throughout the Atlantic.[27] Weather instruments were often taken to sea—as tools of navigation, to warn of impending storms, and to collect observations from far-flung places.[28] But this did not yield any significant knowledge of large-scale atmospheric phenomena. Weather apparatus was also incorporated in the practices of geodesic and cartographic surveying in the eighteenth century. Thermometers and barometers formed part of the equipment of Mason and Dixon, drawing the line between Maryland and Pennsylvania in the 1760s; of William Roy, conducting the Paris–Greenwich triangulation in the 1780s; and of Delambre and Méchain, measuring out the meridian of Paris in the 1790s.[29] But the purpose of these measurements of air temperature, pressure, and humidity was to eliminate their effects on the surveying instruments. While the surveyors did forge a kind of connection between meteorology and geographical knowledge, they were basically trying to exclude the weather from interfering with topographical mapping, rather than to map the atmosphere directly.

As a number of historians have noted recently, instruments were used in innovative ways in the eighteenth century to produce knowledge of geographical space.[30] The accumulation of such knowledge was a significant aspect of scientific development in the period. As Europeans extended their influence worldwide, they deployed techniques of navigation, cartography, natural history, and other sciences to master the spaces they encountered.[31] Primarily, these methods were devoted to charting the oceans and the land—to rendering the seas navigable, physical terrain accountable, and its biological inhabitants classifiable. Insofar as the space of the atmosphere came to be known, however, it was largely by techniques that demanded that the observer be stationary, rather than in motion. Knowledge of the air, the traditional domain of meteorology, remained mostly the preserve of settlers, rather than of travelers. As we have seen in connection with colonial outposts, fairly prolonged residence was required in order to gain knowledge of the climatic characteristics of a place. What was sought was information about the patterns of the seasons and how they related to agriculture and human health. This knowledge demanded observation in a localized spatial domain over an extended chronological period.

This was the most significant accomplishment of eighteenth-century weather science. What was produced was not knowledge of weather systems in the nineteenth-century mode, but rather information about the climates of regions, whether they were provinces, nations, or continents. As we have seen in the British case, the assembled records of the weather contributed collectively to a new consciousness of the national climate, identified with the regular patterns of weather experienced on the island. The concept was put to use in political and historical writing, in medical discourse, and in journalism. It amounted to a geographical expansion of what the term *weather* referred to, broadening it out from something experienced at a single place or in a very limited area to something that could explain the character of a people, their state of health, and aspects of their historical destiny. Although the concept was certainly geographical in a general way, it did not lead to any actual maps of the atmosphere. Talk about a national climate assigned a certain kind of weather to a certain territory, but it did not explain what happened in the atmosphere to cause it. Looking back from the vantage point of the following century, one could say that it did not get to grips with the spatial extent of atmospheric phenomena themselves. These were only properly comprehended when communications technology improved.

From the retrospective point of view, one could certainly deliver a negative judgment on the progress of meteorological science in the eighteenth century. It could be said that the data being accumulated led to no dramatic theoretical advances. A new understanding of the dynamics of the atmosphere was achieved beyond the horizon of the period, in a way that seems discontinuous with the earlier information gathering. That enterprise was proceeding not so much in the wrong direction as in the wrong dimension, collecting information about chronological sequences of weather in particular places, whereas what was needed was an understanding of spatially extended—and rapidly changing—atmospheric formations. But such a judgment would be one-sided and arguably unhistorical. It would ignore what eighteenth-century investigators were trying to accomplish, their own understanding of the imperatives of their inquiry. As we have seen, they catalogued the weather in their own localities as a contribution to an overall knowledge of the climate in which human life was lived. Their accumulating knowledge of the weather participated in a widening geographical awareness, the climate of a nation being seen as an attribute of its geography and defined by contrast with conditions prevailing elsewhere. These observers also saw their inquiry as consistent with historical change as they understood it. Pursuing their task over the course of years, or even decades, they were acting in accordance with a vision of history as continuous progress. They worked to recuperate vernacular knowledge for the purposes of science, exploiting popular traditions in a way that reflected their notions of cultural improvement. As knowledge of the atmosphere steadily accumulated, it was thought, the weather would be civilized—reduced to norms and regular laws, even altered to be more moderate and less threatening—as part of the overall progress of civilization.

In these respects, the science of weather in the eighteenth century was a *reflexive* enterprise, mirroring investigators' awareness of their situation and context. Eighteenth-century scientific practitioners saw themselves as participants in an expanding domain of enlightenment. They expected empirical knowledge to increase hand in hand with material progress. But at the same time, part of the experience of enlightenment was the understanding that the process was neither complete nor unopposed. It was appreciated that scientific rationality confronted many traditional beliefs and customs, and indeed that its powers were limited even when it came to determining the behavior of a single individual. The persistence of traditional weather lore, proverbs, and sayings, the survival of what were viewed as "superstitions" among the populace at large, reminded investigators that their society had not been totally enlightened. Particularly when anomalous or extreme weather occurred, the project of reducing the climate to providential regularity seemed to be thwarted. Such incivilities in the atmosphere tended to evoke incivilities also among the human population. The resurgence of superstitious fears forcefully conveyed to enlightened intellectuals the fact that cultural change in the society around them was only partial. Growing knowledge brought with it an enhanced understanding of how much human beings depended on their emotions and bodily health. Cultural change also raised people's awareness of the diversity that sheltered under the rubric of "human nature," the cultural localism that persisted in the face of attempts to extend knowledge across the globe. In these ways, knowledge of the weather and climate reflected an awareness of the intrinsic limits of enlightenment, the incompleteness that was an inherent part of the movement as it was historically experienced.

Eighteenth-century thinkers on these matters expressed a degree of historical self-consciousness that sometimes eludes us today. They initiated systematic inquiry into the influence of climate at different stages of social development. Pursuing this, they recognized the inextricability of human society and its natural environment, the fact that advancing civilization did not free humanity from its dependence on nature, but rather ramified and deepened the connections between them. It was a lesson reinforced by the attempts of intellectuals to extend the domain of scientific reason over the weather and by the limited scope of their accomplishment. And it is a lesson we might feel is still worth attending to.

As I was finishing this book, in the late summer of 2005, Hurricane Katrina struck the Gulf Coast of the United States with devastating impact on Louisiana and Mississippi. Attending to the news and commentary from New Orleans and elsewhere, I found it hard not to think of parallels to eighteenth-century weather disasters. Some journalists commented on the resurgence of apocalyptic fears among the population affected by the hurricane. Sometimes they treated these beliefs sympathetically. Charles Passy in the *Palm Beach Post* sounded out preachers and environmentalists and found them purveying essentially the same message: "If we are more vulnerable, it's perhaps because we're reaping what we've sown."[32] Deborah Caldwell of *The Advocate* in Baton Rouge noted, similarly, that Christian conservatives and leftist environmentalists were both inclined to view Katrina as a punishment for human transgressions.[33] In a more sophisticated analysis, Edward Rothstein in the *New York Times* remarked on the persistence of a "scientific/moral theodicy in which human sin is still a dominant factor."[34] Some commentators, it was clear, did not so much endorse these beliefs as exhibit them as specimens of resurgent superstitions. The widespread belief that the hurricane constituted a punishment was to be viewed as itself a sign that cultural primitivism was reemerging. This theme complemented the stories about the collapse of social order among some of the population abandoned in New Orleans, though it later turned out that many of these stories were exaggerated. Without denying the specific roots of these reports in contemporary American social anxieties and racial tensions, I could not help noting the resonances with my own studies of the eighteenth century. At that time, as I have shown, unusual or catastrophic weather events focused worries about science, enlightenment, and modernity. The regular climate was viewed as benevolent, a force that integrated human beings with their environment, that answered to their needs, and could even be modified by the progress of civilization. But occasional weather disasters brought out profound doubts about progress, about its limits and its drawbacks. My studies have indicated how some of the cultural responses to Hurricane Katrina have deep roots indeed. The weather apparently still has the power not just to disrupt our material lives, but also to make us reflect on the shallowness of civilization in our incompletely enlightened age.

PREFACE

1. Examples include Boia, *Weather in the Imagination*; and Cathcart, *Rain*.
2. Serres, *Natural Contract*, 27. For more of Serres's philosophical reflections on the weather, see Serres, *Natural Contract*, 3–4, 27–30; *Genesis*, 101–2; and *Birth of Physics*, 67–68; and, for commentary, Bate, "Living with the Weather."
3. Discussions by journalists along these lines include Rothstein, "Seeking Justice"; Seabrook, "Selling the Weather"; and Svenvold, "Look Dear—More Catastrophes!"
4. Kant's 1784 statement on enlightenment is included, with extensive commentary, in Schmidt, *What Is Enlightenment?* See also Clark, Golinski, and Schaffer, "Introduction."

INTRODUCTION   *Weather and Enlightenment*

1. Chapter 1 will give more information about the document and evidence about the author. See also Golinski, " 'Exquisite Atmography.' "
2. On Howard, see Hamblyn, *Invention of Clouds*. The 1783 episode will be discussed in more detail in chapter 2.
3. Howard, *Climate of London*, 2:292.
4. Cathcart, *Rain*; Harley, "Nice Weather." For comparable American attitudes, see Meyer, *Americans and Their Weather*.
5. See especially Glacken, *Traces on the Rhodian Shore*.
6. For examples, see Kington, *Weather of the 1780s*; and Risk Management Solutions, *December 1703 Windstorm*.
7. See especially Feldman, "Late Enlightenment Meteorology." Feldman builds upon an earlier historiography that emphasized the use of instruments to compile observations as characteristic of meteorology in this period. See, for example, Frisinger, *History of Meteorology*; Shaw, *Meteorology in History*; and Wolf, *History of Science*, 1:274–341.
8. Anderson, *Predicting the Weather*.
9. The point is argued eloquently in Outram, "The Enlightenment Our Contemporary."
10. Midgley, *Science and Poetry*, 160.
11. Latour, *We Have Never Been Modern*, esp. 70–71. Also, on postmodern affinities with the Enlightenment, see Baker and Reill, *What's Left of Enlightenment?*
12. For more on this historiographical development, see Clark, Golinski, and Schaffer, "Introduction."

13. Examples include Brewer, *Pleasures of the Imagination*; Brewer and Porter, *Consumption and the World of Goods*; Capp, *Astrology and the Popular Press*; Clark, Golinski, and Schaffer, *Sciences in Enlightened Europe*; Golinski, *Science as Public Culture*; Porter, *Creation of the Modern World*; and Sutton, *Science for a Polite Society*.

14. See the chapters in part 4 of Clark, Golinski, and Schaffer, *Sciences in Enlightened Europe*. Much of the work on the geographical diversity of the Enlightenment was inspired by Porter and Teich, *The Enlightenment in National Context*.

15. Representative of this perspective are Merchant, *Death of Nature*; and Easlea, *Science and Sexual Oppression*.

16. On these interpretations of the Enlightenment, see Clark, Golinski, and Schaffer, "Introduction"; Sluga, "Heidegger and the Critique of Reason"; and Schmidt, *What Is Enlightenment?* 15 – 31, 345 – 67.

17. Particularly informative studies of the mythical resonances of Shelley's work are Baldick, *In Frankenstein's Shadow*; and Bann, *Frankenstein, Creation and Monstrosity*.

18. For an illuminating discussion, see Turney, *Frankenstein's Footsteps*.

19. Foucault, "What Is Enlightenment?" 45. On Foucault's relation to the Enlightenment, see Schmidt and Wartenberg, "Foucault's Enlightenment," 303 – 4: "It was never, for him, a question of deciding 'for' or 'against' the enlightenment—as if we could somehow manage to disavow an event which has, in fundamental ways, defined how we think about ourselves."

CHAPTER ONE   *Experiencing the Weather in 1703: Observation and Feelings*

1. 1703 Weather Diary, consulted in the Lancing College Archive, Lancing, West Sussex (hereafter referred to as "1703 Diary"), 242 (entry dated 10 January 1703). The document was temporarily moved to the West Sussex Record Office, Chichester, West Sussex (accession no. 12,761). There is also a handwritten copy in the National Meteorological Archive, Bracknell, Berkshire. Punctuation of quotations from the diary has been slightly simplified to ease readability. Page numbers are as marked on the pages of the manuscript volume, from 241 to 448, with four unnumbered pages at the end. Page numbers for quotations from the diary in this chapter will henceforth be given in the text. (The phrase "trembling air" is borrowed from Edmund Spenser's poem "Prothalamion.")

2. The diarist was apparently away from home on a visit to Caldwell, near Kidderminster, at the beginning of the year; he began keeping his fuller record on return to his home at Edgiock, Worcestershire.

3. See, for example, Cassidy, "Meteorology in Mannheim"; Feldman, "Late Enlightenment Meteorology"; Frisinger, *History of Meteorology*; Giroux, "Genèse de la météorologie scientifique"; Janković, "Meteorology"; Kington, *Weather of the 1780s*; Shaw, *Meteorology in History*; and Wolf, *History of Science*, 1:274 – 341.

4. See especially Sherman, *Telling Time*.

5. Golinski, "'Exquisite Atmography.'" P. R. Zealley, who transcribed the diary in a series of seven notebooks in the 1920s, published two short notes on the document:

Zealley, "A Florid Weather Diary, Worcestershire, 1703"; and Zealley, "A Florid Weather Diary." He ascribed authorship to John Whiston (1711−80), who, however, was born eight years after the diary was compiled.

6. The inscription reads: "To Mr. Thomas Barker from his Very affectionate & obliged Kinsman, J. Whiston, Sept. 2nd 1746. Ex Bibliotheca Hen. Bland, D.D., Decani Dunelm: & Coll. Eton. Praepositi." On Barker, see Kington, *Weather Journals*; on Bland, see Leigh, *Eton College Register*. Henry Bland (1677/8−1746) was a fellow of King's College, Cambridge, from 1699 to 1702, rector of Great Bircham in Norfolk from 1704, and subsequently headmaster of Eton and dean of Durham; he died on 24 May 1746. He is not known to have had any connection with Worcestershire, where the diary was compiled. Nor is there any mention in the document of the circumstances of Bland's life, such as his marriage on 24 January 1703. For these reasons, it seems most unlikely that he was himself the author. Inquiries at the Eton College archives have revealed no further sources on Bland.

7. Edgiock is mentioned in the diary on, for example, p. 423, where it is said to be "title & Residence to our ancestors." This location is confirmed by references in the diary to nearby places. For example, the Severn valley, Bredon Hill, and Malvern Hills (said to be exactly southwest) are in view; Berrow Hill ("Borough hill" in the diary) is nearby; news is received from nearby Alcester and Ipsley; and the author attends church at Hanbury (about nine kilometers distant). Information about the ownership of the manor house at Edgiock is given in Willis-Bund and Doubleday, *Victoria History of the County of Worcester*, 3:425−26, and in works by local historians of the Inkberrow parish. Bradbrook, *History of the Parish of Inkberrow*, gives summaries of the documents relating to the sale of the manor. Hunt and Jackson, *More about Inkberrow*, adds further details. A drawing of the house, made in 1811, is printed in Hunt and Jackson, *Inkberrow Ways*, 39.

8. This is the main reason I have assumed that the author was male, taken together with the absence of references to domestic circumstances in the diary. The possibility that the diarist might have been a woman is an intriguing one; if demonstrated, it would call for substantial reinterpretation of the document. But at present, there is nothing to support this possibility. Valuable information about women who kept various kinds of technical and philosophical journals can be found in Hunter and Hutton, *Women, Science, and Medicine*. On a female weather diarist of the late eighteenth century, see Wheeler, "Margaret Mackenzie of Delvine, Perthshire."

9. Foster, *Alumni Oxonienses*, s.v. "Appletree." A visit to the Balliol College archives yielded no further information about Thomas Appletree. There is information about the Deddington branch of the family in *Victoria History of the County of Oxford*, 9:92−93. There is no mention of Appletree in Hearne, *Remarks and Collections*; Wood, *Athenae Oxonienses*; or Wood, *Fasti Oxonienses*.

10. Hanbury Parish Register and Will of Thomas Appletree, in Hereford and Worcestershire Record Office. The will was proven at Worcester on 3 August 1728 and executed on 1 February 1729.

11. Ibid. The inventory for the estate sale is dated 14 August 1728. Edward Hollis was to receive £40 and "my wearing Apparrel all but my shirts Cravats & Handkerchiefs," and Anne Hollis was to get the bed and curtains.

12. References to Boyle's work appear in the diary on, for example, pp. 296, 313, 325, 361, 363, 387, and 435.

13. Gunther, *Early Science in Oxford*, 1:43–51; Overnell, *Ashmolean Museum*, 23; Simcock, *Ashmolean Museum*, 7–10; Guerrini, "Chemistry Teaching."

14. Christ Church Library MS 427, f. 21, quoted in Frank, "Medicine," 427.

15. Frank, "Medicine," 432–34.

16. Guerlac, "The Poets' Nitre"; Debus, "The Paracelsian Aerial Niter"; Frank, *Harvey and the Oxford Physiologists*, 117–28, 221–45. Niter was a frequent preoccupation of Oxford chemists for more than a century, since it regularly effloresced from the walls of the laboratory, which had been built on ground contaminated by waste from the Exeter College "privy house." See Simcock, *Ashmolean Museum*, 5, 30 n. 52.

17. Willmoth, "John Flamsteed's Letter."

18. Rossi, *Dark Abyss of Time*, 12–17.

19. Lister, "Three Papers."

20. Bohun, *Discourse*, 23–26.

21. Leigh, *Natural History of Lancashire*, 6, 13; Plot, *Natural History of Staffordshire*, 26.

22. Gunther, *Early Science in Oxford*, 14:283; see also 159, 268, 278.

23. See, for example, Leigh, *Natural History of Lancashire*, 6, 9; Plot, *Natural History of Oxfordshire*, 1–17; and Plot, *Natural History of Staffordshire*, 1–31; and, for a discussion of the tradition as a whole, see Janković, *Reading the Skies*, esp. 78–102.

24. Robinson, *New Observations*, 181.

25. For these episodes, see Suetonius, *The Twelve Caesars*, 174; and 1 Kings 18:36–40.

26. For later instances, see Perkins, *Visions of the Future*, 201–2; and Inwards, *Weather Lore*.

27. Cf. Swainson, *Handbook of Weather Folk-Lore*, 205.

28. On 1 November 1703, conditions were said to be typical of "Allhallen-summer" by a person the diarist designated "my G:F:" On All-hallow summer, see the reference to Shakespeare's *I Henry IV* in Blackburn and Holford-Strevens, *Oxford Companion to the Year*, 441. See also Volney, *View of the Climate and Soil*, 262n, on "All-hallows summer" as the European equivalent of Indian summer in America. The "G:F:" referred to might be Thomas Appletree's grandfather, still living at Edgiock in 1703.

29. Atkinson, "William Derham"; John Locke, "Register of the Air," Bodleian Library, Oxford, MS 48120, ff. 466–531. See also Hunter and Gregory, *Astrological Diary*, 15–16; and Schove and Reynolds, "Weather in Scotland."

30. Webster, "Writing to Redundancy," 40. See also Foucault, "On the Genealogy of Ethics"; and Martin, Gutman, and Hutton, *Technologies of the Self*. Other sources on early-modern diaries include Matthews, *British Diaries*; Macfarlane, *Family Life of Ralph Josselin*, 3–11; Hunter and Gregory, *Astrological Diary*; and Mulligan, "Self-Scrutiny."

31. *Oxford Classical Dictionary*, s.v. "Meteorology"; Taub, *Ancient Meteorology*.

32. 1703 Diary, 345, 388 (quoting Aristotle, *Meteorologica* 3:2). On Aristotle's meteorology, see Taub, *Ancient Meteorology*, 77–115.

33. 1703 Diary, 352, 370, 373–74. On the ancient writers, see Nielsen and Solomon, "Writing the Bodies of Water"; and Sarasohn, "Epicureanism."

34. Taub, *Ancient Meteorology*, 127–41. On 13 October 1703, the diarist recorded that the placid weather "infused nothing but a divine calm & ἀταραξία [*ataraxia*]" (402).

35. Seneca, *Natural Questions*, 1:197 (translation by Thomas H. Corcoran). The diarist quotes a passage from slightly earlier in the text, which he appears to mistranslate: "Malo fulmina non timere quam nosse, the way to Contemn it is to know it" (388). In fact, the statement is made by the interlocutor, rather than in the voice of the author, in Seneca's text; it is translated by Corcoran as "I should rather I did not fear lightning than know about it" (1:193).

36. A few days later, the author recorded a night rhapsody in which he imagined himself transported among the stars, "winged by Philosophic curiosity to dive into ye stellar deep celestiall abyss, & penetrate ye intimate recesses, ... & there in raptures descry millions of worlds wich undiscovered lie as yet to Lyncean Telescope or eie" (359). Such astronomical fantasy could draw on a distinguished literary inheritance from the seventeenth century, including Kepler's *Somnium* (1634). For a recent discussion, see Campbell, *Wonder and Science*, 133–43.

37. Ixion was the legendary king who tried to rape the goddess Hera but was deceived by an image of her fabricated from clouds by her husband, Zeus. So, the clouds were not the womb of Ixion, as the diarist states, but rather the objects toward which his lust was diverted.

38. On the contemporary critique of "enthusiasm" as a pathological susceptibility to material influences, rather than a genuinely spiritual experience, see Heyd, "Medical Discourse." For more on religious enthusiasm and how it was explained in terms of vapors and the imagination, see Vermeir, "'Physical Prophet.'"

39. Robinson, *New Observations*, 109.

40. On Kepler's "world soul," see Simon, *Kepler, astronome astrologue*, 44–48. I thank Nicholas Jardine for help with this point.

41. On this topic, see the essays collected in Lawrence and Shapin, *Science Incarnate*. See also Daston and Galison, "Image of Objectivity"; and Golinski, "Care of the Self."

42. More, *Enthusiasmus Triumphatus*, 5; [Trenchard], *Natural History of Superstition*, 14.

43. Burton, *Anatomy of Melancholy*, 1:379–81. A more systematic investigation would probably uncover many echoes of Burton's work in the 1703 diary.

44. Serres, *Birth of Physics*, 68.

45. Hamblyn, *Invention of Clouds*.

46. Townley, *Journal Kept in the Isle of Man*; Corbin, *Lure of the Sea*, 90–95 (quotation on 91).

47. Castle, "Female Thermometer."

CHAPTER TWO *Public Weather and the Culture of Enlightenment*

1. [Addison et al.], *The Spectator*, 1:256 (no. 68, 18 May 1711).

2. Modern accounts of the storm rely ultimately on [Defoe], *The Storm*. See also Brayne, *Greatest Storm*; Risk Management Solutions, *December 1703 Windstorm*; Janković, *Reading the Skies*, 59–64; and Hamblyn, "Introduction."

3. Evelyn, *Diary and Correspondence*, 531.

4. *Wonderful History*, 53.

5. Gifford, *Sermon*, 24.

6. [Defoe], *The Storm*. Notwithstanding the notorious difficulties of Defoe's bibliography, the attribution of this work to him seems secure, according to Furbank and Owens, *Critical Bibliography of Daniel Defoe*, 54−56. Some of its contents overlap with *An Elegy on the Author of the True-Born-English-Man* (first published in 1704), of which Defoe is confidently identified as the author. See also Hamblyn, "Introduction."

7. *Exact Relation of the Late Dreadful Tempest*, 3.

8. *Terrible Stormy Wind*, 8−10. Other commentaries include *Amazing Tempest*; *Wonderful History*, 29−53; Bradbury, *God's Empire over the Wind*; Hussey, *Warning from the Winds*; and Gifford, *Sermon*.

9. [Defoe], *The Storm*, A8r, 2.

10. This Richard Townley should not be confused with the author of the *Journal Kept in the Isle of Man*, who lived nearly a century later.

11. Derham, "A Letter . . . Containing His Observations concerning the Late Storm," 1531, 1532. See also Derham, *Physico-Theology*, 14−19, on the providential importance of winds for human health and prosperity.

12. Leeuwenhoek, "Part of a Letter . . . Giving His Observations on the Late Storm," 1537; Fuller, "Part of a Letter . . . Concerning a Strange Effect of the Late Great Storm," 1530.

13. A decade later, the author of the anonymous *Essay concerning the Late Apparition in the Heavens* professed disappointment that intellectuals would say among themselves that such phenomena were natural but were reluctant to do so publicly. Perhaps ironically, he suggested that this reluctance itself increased popular fear of atmospheric anomalies. (*Essay concerning the Late Apparition in the Heavens*, 1−2)

14. "S. W.", *A Poem on the Late Violent Storm*, 1.

15. "W. F." [William Fulke], *Meteors*, 7, 33.

16. On contemporary ideas about providential interventions in politics, see Spurr, " 'Virtue, Religion and Government' "; Worden, "Providence and Politics"; Reedy, "Mystical Politics"; Israel and Parker, "Of Providence and Protestant Winds"; and Schechner, *Comets*, 66−88.

17. *Practical Discourse on the Late Earthquakes*, 15. On the tradition of meteors as wonders, see also Janković, *Reading the Skies*; and Heninger, *Handbook of Renaissance Meteorology*.

18. Cook, *Edmond Halley*, 351−53.

19. Fara, "Lord Derwentwater's Lights"; Janković, *Reading the Skies*, 68−77.

20. Cook, *Edmond Halley*, 347; Wolf, *History of Science*, 1:303−5.

21. Burns, *Age of Wonders*, 152−64. See also Whiston, *Account of a Surprizing Meteor*.

22. Burns, " 'Our Lot Is Fallen' "; Burns, *Age of Wonders*; Thomas, *Religion and the Decline of Magic*, 90−132; Daston, "Marvelous Facts."

23. On winds, see Bohun, *Discourse*.

24. Rossi, *Dark Abyss of Time*, 25−49.

25. *Wonderful History*, unnumbered pages.

26. *True and Particular Account of a Storm*, 15−16.

27. Pointer, *Rational Account of the Weather*, vii.

28. Ibid., 195.

29. Ibid., 197, 196, 199.

30. Daston and Park, *Wonders*, 329−63; Burns, *Age of Wonders*; Worden, "Providence and Politics."

31. Daston, "Marvelous Facts."

32. Janković, *Reading the Skies*, 53.

33. Gifford, *Sermon*, 8−10.

34. *True and Particular Account of a Storm*, 21. One of the moralizing commentators on the 1703 storm noted similarly, "Tho' things have but a Cloudy Face at Present, and perhaps there's not much appearance of Wisdom in 'em, yet we may conclude, that God Orders them all aright." (Bradbury, *God's Empire over the Wind*, 21)

35. Budgen, *Passage of the Hurricane*, 11.

36. [Defoe], *The Storm*, 33.

37. On the "strange but true," see McKeon, *Origins of the English Novel*, 65−89; and Shapiro, *Culture of Fact*, 86−104.

38. Bradbury, *God's Empire over the Wind*, 31.

39. The "public sphere" has been the topic of extensive commentary since Habermas, *Structural Transformation of the Public Sphere*. On the connection to the uniformity of nature, see Clark, Golinski, and Schaffer, "Introduction," 16−26.

40. Pointer, *Rational Account of the Weather*, iii.

41. An early example is *A Diary or Weather-Journall* [c. 1685], published by the London instrument maker John Warner (British Library call number 816.m.7(103)).

42. Locke, quoted in Dewhurst, *John Locke*, 301.

43. Rusnock, *Correspondence of James Jurin*, 27−31; Rusnock, *Vital Accounts*, 110−16.

44. Feldman, "Late Enlightenment Meteorology."

45. Sherbo, "English Weather"; Fothergill, *Works*, 77−128. See also Porter, "Lay Medical Knowledge."

46. On normalization, see Serres, *Natural Contract*, 4.

47. See especially Colley, *Britons*; and Wilson, *Island Race*.

48. [Addison et al.], *The Spectator*, 1:262 (no. 69, 19 May 1711).

49. Swift, *Gulliver's Travels*, 167.

50. Campbell, *Political Survey of Britain*, 1:47. See also Williams, *Climate of Great Britain*, 2−3.

51. Campbell, *Political Survey of Britain*, 1:51, 54, 55.

52. English commentators, on the other hand, sometimes attributed insular status to their own country, conflating England with the island of Britain as a whole. See, for example, Williams, *Climate of Great Britain*, 2.

53. Dixon, "Early Irish Weather Records."

54. Schove and Reynolds, "Weather in Scotland"; Hay, *Diary*.

55. Wheeler, "Weather Diary of Margaret Mackenzie"; Wheeler, "Margaret Mackenzie of Delvine, Perthshire." The second article includes some remarks on other Scottish women who kept weather diaries. A slightly later female observer in England was Caroline Molesworth, who kept her diary from 1825 to 1850 at Cobham, Surrey. See Molesworth, *Cobham Journals*.

56. Rutty, *Essay towards a Natural History of the County of Dublin*, 2:275, 280−81.

57. Hillary, *Practical Essay on the Small-Pox*, 40; Wintringham, *Commentarius Nosologicus*, 252.

58. Pointer, *Rational Account of the Weather*, 87.

59. Wood, *Valetudinarian's Companion*, 19. See also Corbin, *Lure of the Sea*.

60. [Bisset], *Essay on the Medical Constitution of Great Britain*.

61. Fothergill, *Works*, 85.

62. *Lady's Magazine* 17 (1786): 680, quoted in Hamblyn, *Invention of Clouds*, 72.

63. Gidal, "Civic Melancholy."

64. Fothergill, *Works*, 89, 96.

65. Arbuthnot, *Essay concerning the Effects of Air*, 151.

66. Falconer, *Remarks on the Influence of Climate*, 50, 71, 73.

67. In 1667, Thomas Sprat had explained the national interest in experimental philosophy by invoking "the position of our climate, the air, the influence of the heaven, [and] the composition of the English blood." These gave "a good sign, that Nature will reveal more of its secrets to the English, than to others." (Sprat, *History of the Royal Society*, 114−15)

68. Short, *General Chronological History*, 1:viii.

69. Rutty, *Essay towards a Natural History of the County of Dublin*, 2:417, 486.

70. Samuel Say, "A Journal of the Weather at Lostaff [Lowestoft] in Suffolk, from 1695 to 1724, by the Rev. Mr. Say," Bodleian Library, Oxford, MS 35448, 1; Marshall, *Experiments and Observations*, 164.

71. White, *Natural History and Antiquities of Selborne*, 203.

72. Gilbert White, "The Naturalist's Journal," British Library, Additional MSS 31,846, fol. 2r.

73. White, *Natural History and Antiquities of Selborne*, 298; Henry White, quoted in Mabey, *Gilbert White*, 190.

74. Grattan and Brayshay, "Amazing and Portentous Summer."

75. White, *Natural History and Antiquities of Selborne*, 298.

76. Grattan and Brayshay, "Amazing and Portentous Summer"; Kington, *Weather of the 1780s*.

77. Hamilton, "Account of the Earthquakes." The summer haze is also mentioned in Barker, "Abstract of a Register"; and Cullum, "Account of a Remarkable Frost."

78. Franklin, "Meteorological Imaginations." The idea was dismissed by Kirwan, *Of the Variations of the Atmosphere*, 219.

79. *Jackson's Oxford Journal*, 12 July 1783, 1, quoted in Hamblyn, *Invention of Clouds*, 71.

80. Burke, *Popular Culture in Early Modern Europe*, 207−86; Malcolmson, *Popular Recreations*, 89−171; Barker-Benfield, *Culture of Sensibility*; Payne, "Elite versus Popular Mentality."

81. [Addison et al.], *The Spectator*, 1:39 (no. 10, 12 March 1711); Hume, *Essays: Moral, Political, and Literary*, 533−37.

82. On conversation and politeness, see Burke, *Art of Conversation*; Klein, "Coffeehouse Civility"; Klein, "Enlightenment as Conversation"; Brewer, *Pleasures of the Imagination*, 34−39, 98−113; and Carter, *Men and the Emergence of Polite Soci-*

ety, 60–70. On women and popular science, see Mullan, "Gendered Knowledge, Gendered Minds"; Douglas, "Popular Science"; Shteir, " 'Conversable Rather Than Scientific' "; and Walters, "Conversation Pieces."

83. Such manuals of polite conversation included *The Art of Complaisance*; [Forrester], *Polite Philosopher*; and [Constable], *Conversation of Gentlemen*.

84. [D'Ancourt], *Lady's Preceptor*, 49.

85. Adams, *Short Dissertation on the Barometer*, 1.

86. Johnson, *The Idler*, no. 11 (24 June 1758), in Johnson, *Works*, 2:36.

87. Inwards, *Weather Lore*, v.

88. Richardson, *Clarissa*, 7:88.

89. There are interesting comments on English people's conversations about the weather in Cathcart, *Rain*, 51–54, where Wilde is quoted on p. 54. See also Myer, "Talking of the Weather." Tobias Smollett satirized the pedantic individual who took the weather too seriously as a topic of conversation in a scene in his *Adventures of Peregrine Pickle* (1784): "The conversation first turned upon the weather, which was investigated in a very philosophical manner by one of the company, who seemed to have consulted all the barometers and thermometers that ever were invented, before he would venture to affirm that it was a chill morning." (662)

90. Johnson, *The Idler*, no. 11 (24 June 1758), in Johnson, *Works*, 2:37.

91. *The Mirror*, no. 35 (1779), quoted in *Oxford English Dictionary*, s.v. "Weather": "The conversation began about the weather, my aunt observing that the seasons were wonderfully altered in her memory." See also Williams, *Climate of Great Britain*, 3. Williams, writing in 1806, declared that the notion that the climate had changed was "an opinion universally adopted of late years."

92. Denham, *Collection of Proverbs*; Swainson, *Handbook of Weather Folk-Lore*; Inwards, *Weather Lore*. See also Garriott, *Weather Folk-Lore*; Shields, "Popular Weather Lore"; and Sloane, *Folklore of American Weather*. Most of the literature on weather proverbs is devoted to collecting them fairly indiscriminately, without any attempt to reconstruct how they were used in everyday life. But for an attempt at a social analysis, see Fox, *Oral and Literate Culture*, 154–57.

93. Inwards, *Weather Lore*, 119–22.

94. For these two, favorite saws of the now-retired BBC weather forecaster Michael Fish, see Denham, *Collection of Proverbs*, 31, 48.

95. Barbara Herrnstein Smith, quoted in Shapin, "Proverbial Economies," 737. For more on how weather proverbs were and are used, see Arora, "Weather Proverbs"; Dundes, "Weather 'Proverbs' "; Ward, "Weather Signs"; and Widdowson, "Form and Function."

96. Davis, "Proverbial Wisdom"; Matthews, "Polite Speech"; Obelkevich, "Proverbs and Social History."

97. Swift, *Swift's Polite Conversation*; Ray, *Collection of English Proverbs*; Fuller, *Gnomologia*.

98. *Knowledge of Things Unknown*.

99. The first edition was Claridge, *The Shepheards' Legacy*. Campbell's role in the 1744 edition has not been generally noticed, though it is mentioned in the *English Short*

Title Catalogue and in the article on Campbell in the old *Dictionary of National Biography*. There is no allusion to the connection in the new *Oxford Dictionary of National Biography*. The 1744 edition was the basis for numerous others, published in London, Oxford, Dublin, and Edinburgh in the course of the eighteenth century. On this, see Janković, *Reading the Skies*, 133−35.

100. Claridge, *Shepherd of Banbury's Rules*, ii.
101. Ibid., iii, viii, 23, 52.
102. Mills, *Essay on the Weather*; Janković, *Reading the Skies*, 139−40.
103. Franklin, "Meteorological Imaginations"; Ray, *Collection of English Proverbs*, 48.
104. Howard, *Climate of London*, 2:vi.
105. Howard, *Seven Lectures on Meteorology*, 1.
106. Howard, *Climate of London*, 2:161.
107. Ibid., 2:198.
108. Howard, *Climate of London*, 1:xxxiv−xxxv.
109. Ibid., vol. 1, unnumbered pages following table 6.
110. On this dismal summer, ascribed by modern researchers to the effects of a volcanic eruption on the island of Sumbawa in the East Indies, see Clubbe, "Tempest-Toss'd Summer of 1816"; and Stommel and Stommel, "Year without a Summer."
111. Howard, *Climate of London*, 1:xxxvi.
112. Marshall, *Experiments and Observations*, 143.

CHAPTER THREE   *Recording and Forecasting*

1. Sherman, *Telling Time*, 1−108; Thompson, "Time, Work-Discipline and Industrial Capitalism."
2. Poole, *Time's Alteration*; Wilcox, *Measure of Times Past*; Manuel, *Isaac Newton*.
3. Serres, *Genesis*, 101−2; Serres, *Birth of Physics*, 67−68. The distinction between *kairos* and *chronos* is discussed by Sherman (*Telling Time*, 10−12), drawing on writings by Frank Kermode and Walter Benjamin, among others.
4. 1703 Diary, 404.
5. Thoresby, quoted in Manley, "The Weather and Diseases," 300; Wintringham, *Commentarius Nosologicus*, 172.
6. Remarks on the weather remain, of course, a feature of personal diaries to this day. For a collection of such remarks, see Head, *Weather Calendar*.
7. Macfarlane, *Diary of Ralph Josselin*; Macfarlane, *Family Life of Ralph Josselin*, 71−75, 188−92.
8. Hay, *Diary*; Schove and Reynolds, "Weather in Scotland."
9. Sources on spiritual journals and their relation to later practices of diary-writing include Webster, "Writing to Redundancy"; Foucault, "On the Genealogy of Ethics"; and Martin, Gutman, and Hutton, *Technologies of the Self*. For suggestions about the broader issue of the place of such documents in the formation of the modern sense of self, see Smith, "Self-Reflection and the Self."
10. Wilson, *Magical Universe*, 51−87; Thomas, *Religion and the Decline of Magic*, 58−89.
11. Burton, *Anatomy of Melancholy*, 3:395, 1:241.

12. MacDonald, *Mystical Bedlam*, 223–25.

13. Beier, *Sufferers and Healers*, 139–53, 159–70, 182–210.

14. Mulligan, "Self-Scrutiny." On Hooke and his diary, see also Shapin, "Who Was Robert Hooke?" and Jardine, *Curious Life of Robert Hooke*, 214–34.

15. Golinski, "Care of the Self."

16. Hooke, "Method for Making a History of the Weather," 175.

17. Nussbaum, *The Autobiographical Subject*, 1–57; Smith, "Self-Reflection and the Self."

18. In this respect, they exemplified the modern ideal of objectivity that has been identified in various areas of scientific practice in this period. For studies of the history of objectivity, see Daston and Galison, "Image of Objectivity"; Daston, "Attention and the Values of Nature"; Megill, *Rethinking Objectivity*; and Shapin, "The Philosopher and the Chicken."

19. *Dictionary of National Biography*, s.v. "Rutty, John"; Harvey, *John Rutty of Dublin*.

20. Rutty, *Spiritual Diary*, 1:x.

21. Ibid., 1:166, 242, 237.

22. Ibid., 1:xi. A pencil note in the British Library copy (call number 4920.b.48) ascribes the introduction to Thomas Hartley.

23. Boswell, *Life of Johnson*, 852–53.

24. Harvey, *John Rutty of Dublin*, 22.

25. Kington, *Weather Journals*, 3–17.

26. Barker, "Account of an Extraordinary Meteor"; Barker, "Account of a Remarkable Halo." The latter phenomenon was witnessed in 1737 but reported nearly twenty-four years later.

27. Kington, *Weather Journals*, 10.

28. "Notebook by Thomas Barker," Houghton Library, Harvard University, bMS Eng. 737, 41, 231. This notebook, compiled at Lyndon from January 1730 to 4 October 1801, contains dates of frosts, astronomical tables, and a diary of natural history from 1736 to 1801. It seems to be the document recorded as missing by Kington in *Weather Journals*, 8. Barker's experiments on very long-term natural changes are paralleled also by his interest in how the colors of stars had changed since antiquity. See Barker, "Remarks on the Mutations of Stars."

29. On Darwin's observations, see Gould, "Worm for a Century."

30. Barker, *Account of the Discoveries*. This book was published by John Whiston, son of the famous astronomer and Barker's uncle.

31. John Locke recorded weather observations at least daily during his residence at Oates in Essex from 1691 to 1703, but it is known that he often spent days in London on business during this period. Presumably, then, he had the record kept by a servant. On Locke's life at this time, see Cranston, *John Locke*, 342–75. For general comments on "invisible technicians," see Shapin, *Social History of Truth*, 355–407; on the invisible aid often given by women family members, see Fara, *Pandora's Breeches*, 109–85.

32. Atkinson, "William Derham," 378.

33. Wintringham, *Commentarius Nosologicus*, 252. Compare Weather Diary of Dr. John Bayly, Chichester, 1769–1773, National Meteorological Archive, 1:35.

34. Observations of Thomas Hughes, Physician, from 1 January 1771 to 17 April 1813, at Stroud, Gloucestershire, National Meteorological Archive.

35. William Borlase kept up his weather diary until the day before his death in 1772, according to Oliver, "William Borlase's Contribution," 276. Joseph Tucker of Rye, Sussex, made his last annotation on 15 October 1733 and died within a few days. See Observations by Joseph Tucker, 1730–1733, at Rye, Sussex, National Meteorological Archive.

36. Strauss and Orlove, "Up in the Air"; Orlove, "How People Name Seasons."

37. Slatkin, "Measuring Authority," 45.

38. On the *parapēgmata*, see Taub, *Ancient Meteorology*, 20–37.

39. Wilson, *Magical Universe*, 51–54.

40. Denham, *Collection of Proverbs*; Swainson, *Handbook of Weather Folk-Lore*; Inwards, *Weather Lore*. For evidence of the survival of such beliefs into the twentieth century, see Pickering, "Four Angels."

41. Blackburn and Holford-Strevens, *Oxford Companion to the Year*.

42. Swainson, *Handbook of Weather Folk-Lore*, 115–16.

43. On Candlemas Day, see Denham, *Collection of Proverbs*, 28; Swainson, *Handbook of Weather Folk-Lore*, 42–50; and Inwards, *Weather Lore*, 20.

44. On the calendar and rural customs, see Malcolmson, *Popular Recreations*, 15–33; Bushaway, *By Rite*, 34–63; and Thomas, *Religion and the Decline of Magic*, 735–45. On moveable feasts as prognostic days, see Swainson, *Handbook of Weather Folk-Lore*, 67–76; and Inwards, *Weather Lore*, 47–48. Seasonal weather lore in France is discussed in Dufour, *Météorologie*; and Galtier, *Météorologie populaire*.

45. Borrowed days are discussed in Swainson, *Handbook of Weather Folk-Lore*, 65–66.

46. Blackburn and Holford-Strevens, *Oxford Companion to the Year*, 144; McWilliams, "Kingdom of the Air," 121.

47. Thomas, *Religion and the Decline of Magic*, 738.

48. Poole, "'Give Us Our Eleven Days!'"; Poole, *Time's Alteration*. On the continuing observance of Old Style dates, see Poole, *Time's Alteration*, 137, 152–57; and Townley, *Journal Kept in the Isle of Man*, 16. Williams, *Climate of Great Britain*, 3–4, records that even in the early nineteenth century, uneducated people or those "addicted to superstition" blamed the calendar change for "the cloudy and ungenial weather we have more or less experienced ever since, and the years of scarcity we have so frequently felt."

49. Examples include *A Diary or Weather-Journall* (c. 1685); and the sheets printed by Daines Barrington and used by Gilbert White almost a century later. See White, "The Naturalist's Journal," British Library, Additional MSS 31,846–31,851.

50. Ray, *Collection of English Proverbs*, 51.

51. Mills, *Essay on Weather*, 72.

52. Inwards, *Weather Lore*, 5.

53. Howard, *Climate of London*, 2:198.

54. Forster, *Perennial Calendar*, xix–xx. On the persistence of seasonal customs and beliefs, see also Webb, "Representations of the Seasons."

55. Perkins, quoted in Poole, *Time's Alteration*, 22.

56. Spadafora, *The Idea of Progress*.

57. Hillary, *Account of the Principle Variations*, x.

58. Wilcox, *Measure of Times Past*.

59. Short, *General Chronological History*.

60. Stillingfleet, *Calendar of Flora*; [Aikin,] *Calendar of Nature*; Webb, "Representations of the Seasons."

61. Claridge, *Shepherd of Banbury's Rules*, 23, 14–15, 46–47.

62. Whewell, *Astronomy and General Physics*, 21–33, 55–62 (quotation on 58).

63. Freud, *Civilization, Society and Religion*, 282 (the passage is from *Civilization and Its Discontents* [1930]).

64. Taub, *Ancient Meteorology*, 29–37, 62–65.

65. Thomas, *Religion and the Decline of Magic*, 335–82, 414–24.

66. Cock, *Meteorologia*, 8–9, 42–43.

67. Curry, *Prophecy and Power*, 57–78; Capp, *Astrology and the Popular Press*, 180–90; Geneva, *Astrology and the Seventeenth-Century Mind*, 75–150.

68. Goad, *Astro-Meteorologia*, 11–12.

69. Flamsteed, quoted in Curry, *Prophecy and Power*, 71.

70. Plot, "Letter . . . to Dr. Martin Lister," 930–31.

71. Boyle, *General History of the Air*, 67–81 (quotation on 76).

72. Curry, *Prophecy and Power*, 95–117; Capp, *Astrology and the Popular Press*, 238–69.

73. Partridge, *Merlinus Redivivus* [1685]. (I have omitted from the quotation the symbols for the moon's phases and aspects of the planets.)

74. Perkins, *Visions of the Future*, 216.

75. Capp, *Astrology and the Popular Press*, 427n. 57; for comment, see Saul, *Historical and Philosophical Account of the Barometer*, 96. On Beighton's weather recording, see Stewart, *Rise of Public Science*, 247–50; on the *Ladies' Diary*, see Costa, "The Ladies' Diary."

76. Andrews, *Royal Almanack*. In 1677, Robert Plot had called for the compilation of "old Almanacks . . . instead of new; that instead of the conjectures of the weather to come, . . . would give a faithful account . . . of the whole weather of the years past, on every day of the month." (Plot, *Natural History of Oxfordshire*, 6)

77. Goad, *Astro-Meteorologica*, 16–17; Hall, *Observations on the Weather*; Hutchinson, *Calendar of the Weather*, 5.

78. Howard, *Climate of London*, 2:227–28.

79. The fact that the eighteen-year period corresponded with a combined lunar-solar cycle was mentioned in Howard, *Climate of London*, 2:277; but the astronomical connection was not specifically avowed in Howard, *Cycle of Eighteen Years*. In the latter work, Howard presented his examination of the cycle as entirely inspired by empirical observations of the weather, without reference to astronomy.

80. Howard, *Seven Lectures on Meteorology*, 1.

81. Curry, *Confusion of Prophets*, 75–76.

82. Anderson, "The Weather Prophets"; Herschel, "The Weather and Weather Prophets."

83. Kirwan, *Estimate of the Temperature*, v.

CHAPTER FOUR *Barometers of Enlightenment*

1. Walters, "Conversation Pieces," provides a good introduction to the material culture of "polite science" in Enlightenment Britain.

2. Bourguet, Licoppe, and Sibum, "Introduction," 7.

3. Outram, "On Being Perseus"; Livingstone, *Putting Science in Its Place*, 72–81.

4. On the social dimension of the process by which instruments are made, see Golinski, *Making Natural Knowledge*, 133–45.

5. Sterne, *Life and Opinions of Tristram Shandy*, 96–98. The classical reference is to Lucian's story of Momus, who rebuked Hephaestus for having made a man without inserting a glass window in his breast by which his heart and passions might be viewed. I am grateful to Otniel Dror, "The Scientific Image of Emotion," for this background.

6. Phelps, *Human Barometer*, 14–15.

7. Debus, "Key to Two Worlds."

8. London broadside, "A Table Plainly Teaching ye Making and Use of a Wetherglas" (1631), reproduced in Godwin, *Robert Fludd*, 60–61.

9. Fludd, *Mosaicall Philosophy* (1638), quoted in Debus, "Key to Two Worlds," 121.

10. Willsford, *Nature's Secrets*, 153, sig. A7v.

11. [Oldenburg and Beale], "Observations Continued upon the Barometer," 164. In 1664, however, Lord Brouncker, the president of the Royal Society, does not seem to have appreciated the distinction, at least to judge from his remarks at a meeting in November of that year, reported to Boyle by Oldenburg. (Oldenburg to Robert Boyle, 17 November 1664, in Oldenburg, *Correspondence*, 2:309–13.)

12. Boyle, *New Experiments and Observations Touching Cold* (1665), in *Works*, 2:485.

13. For example, in the anonymous *Companion to the Weather-Glass* (1796).

14. Smart, "Jubilate Agno," 74.

15. See Middleton, *History of the Thermometer*; Middleton, *History of the Barometer*; Taylor, "Origin of the Thermometer"; and Barnett, "Development of Thermometry."

16. Boyle, *New Experiments and Observations Touching Cold* (1665), in *Works*, 2:487; Middleton, *History of the Barometer*, 71–72.

17. Birch, *History of the Royal Society*, 1:301–2.

18. Oldenburg to Boyle, 19 December 1665, in Oldenburg, *Correspondence*, 2:645–49.

19. Birch, *History of the Royal Society*, 1:304.

20. Oldenburg to Boyle, 30 December 1665, in Oldenburg, *Correspondence*, 2:652–57.

21. Boyle, "Some Observations and Directions about the Barometer," 183–85.

22. Birch, *History of the Royal Society*, 3:460–63; Lister, "Discourse concerning the Rising and Falling of the Quicksilver." A view comparable to Lister's was voiced by Charles Leigh in 1700, in his *Natural History of Lancashire*, 10.

23. Garden, "Discourse concerning Weather"; Wallis, "Discourse concerning the Air's Gravity."

24. Halley, "Discourse of the Rule of the Decrease," 110–11.

25. Halley, "Historical Account of the Trade Winds."

26. Albury, "Halley and the Barometer," 223.

27. Smith, *Compleat Discourse*, 79–93; Smith, *Horological Disquisitions*, 73–77; Neve, *Baroscopologia*, 28–29.

28. Harris, *Lexicon Technicum*, s.v. "Barometer"; Desaguliers, *Lectures of Experimental Philosophy*, 135–36; Saul, *Historical and Philosophical Account of the Barometer*, 33–38.

29. Smith, *Horological Disquisitions*, 89; Parker, *Account of a Portable Barometer*, 84.

30. [Hale], *Difficiles Nugae*. For comments by a contemporary, see North, *Notes of Me*, 167–68.

31. Bryden, "Balance Barometer," 364.

32. Testimonial by John Patrick, 13 April 1727, attached to a letter by Johann von Hatzfeld, Royal Society of London Early Letters, H3:127. I thank Simon Schaffer for this reference.

33. Roubais du Tourcoin, *Physical Dissertation*, 7.

34. Levere and Turner, *Discussing Chemistry and Steam*, 87–88.

35. Kirwan, "Essay on the Variations of the Barometer."

36. Ibid., 44.

37. The concept of the "trading zone" is developed in Galison, *Image and Logic*.

38. Roger North, Life of Francis North (1742), quoted in Goodison, *English Barometers*, 31; Saul, *Historical and Philosophical Account of the Barometer*, 1.

39. Neve, *Baroscopologia*, 4–5.

40. Roubais du Tourcoin, *Physical Dissertation*, 15.

41. Roger North, "Essay of the Barometer," British Library, Additional MSS 32,541, f. 17r–v. See also North, *Notes of Me*, 201; and Middleton, *History of the Barometer*, 100–101.

42. Boyle, *General History of the Air*, 141–42. On wheel barometers, see Middleton, *History of the Barometer*, 94–99.

43. Middleton, *History of the Barometer*; Goodison, *English Barometers*, 30–41; Taylor, *Mathematical Practitioners*. I am grateful also for information supplied from the Project Simon database of early-modern instrument makers by Dr. Gloria C. Clifton of the National Maritime Museum.

44. Harris, *Lexicon Technicum*, s.v. "Barometer."

45. Patrick, *New Improvement*, 1.

46. Castle, "Female Thermometer," 22.

47. Sinclair, *Principles of Astronomy and Navigation*, 48.

48. The information about barometer makers is from Goodison, *English Barometers*; Banfield, *Barometer Makers and Retailers*; and Crawforth, "Evidence from Trade Cards."

49. On changes in barometer design, see Banfield, *Antique Barometers*. On consumer culture in general, see Brewer and Porter, *Consumption and the World of Goods*; and Brewer, *Pleasures of the Imagination*.

50. Patrick, *New Improvement*, 1.

51. Neve, *Baroscopologia*, 3.

52. Martin, *Description of the … Simple Barometer*, 3.

53. Laurence, *Fruit-Garden Kalendar*, 135.

54. Oliver, "William Borlase's Contribution," 279–81.

55. White, *Natural History and Antiquities of Selborne*, 282–83.

56. Kington, *Weather Journals*, 26–27.

57. Howard, *Climate of London*, 2:145.

58. Smith, *Compleat Discourse*, 1–49; Neve, *Baroscopologia*, 4–8.

59. "To the Authors of the *Universal Magazine*"; Walters, "Tools of Enlightenment," 120–21; *Diary or Weather-Journall*.

60. Saul, *Historical and Philosophical Account of the Barometer*, 1.

61. Ibid.

62. The painting is shown in Raines, *Marcellus Laroon*, plate 54. I have not been able to locate the original, which is believed to be in a private collection.

63. Martin, *Young Gentleman and Lady's Philosophy*, 1:323, 324.

64. Saul, *Historical and Philosophical Account of the Barometer*, 2.

65. Martin, *Young Gentleman and Lady's Philosophy*, 1:324–25, 328.

66. North, "Essay of the Barometer," f. 17v.

67. Johnson, *The Idler*, no. 17 (5 August 1758), in Johnson, *Works*, 2:53–54.

68. Johnson, *The Idler*, no. 33 (2 December 1758), in Johnson, *Works*, 2:101–6; cf. Castle, "Female Thermometer," 13–14.

69. North, "Essay of the Barometer," f. 14v.

70. Ibid., f. 18v.

71. Neve, *Baroscopologia*, 33.

72. Goodison, *English Barometers*, 32; Capp, *Astrology and the Popular Press*, 185, 203.

73. Parker, *Account of a Portable Barometer*.

74. [Parker], *New Account of the Alterations of Wind and Weather*; [Parker], *New Baroscopical Account*.

75. [Gadbury], *Stars and Planets*, 7. Gustavus Parker should not be confused with the more famous George Parker, author of a popular astrological almanac.

76. Ibid., unnumbered page in preface.

77. [Parker and Patrick], *New Account of the Alterations of the Wind and Weather*.

78. North, "Essay of the Barometer," f. 294r–v.

79. Moore, *Vox Stellarum* (1791), 36–37.

80. Miller, *Gardener's Dictionary*, s.v. "Barometer."

81. Peter Rabalio, "Rules for Foretelling the Alteration of the Weather by the Barometer," copy in the Whipple Museum of the History of Science, Cambridge.

82. Claridge, *Shepherd of Banbury's Rules*, vii.

83. Ibid., iv.

84. Smith, *Horological Disquisitions*, 78–88.

85. Laurence, *Fruit-Garden Kalendar*, 144–49.

86. Mills, *Essay on the Weather*, 74.

87. Laurence, *Fruit-Garden Kalendar*, 141.

88. Martin, *Description of the . . . Simple Barometer*, i–ii. See also Pointer, *Rational Account of the Weather*, xxiv; and Magellan, *Description et usage des nouveaux baromètres*, 140–41.

89. Adams, *Short Dissertation on the Barometer*, iv.

90. Ibid., iv, 27–28.

91. Short, *New Observations*, 456.
92. North, "Essay of the Barometer," f. 20v.
93. *Meteorologist's Assistant*, 2.

CHAPTER FIVE   *Sensibility and Climatic Pathology*

1. See especially Barker-Benfield, *Culture of Sensibility*.
2. Ulrich von Wilamowitz-Moellendorff, quoted in *Oxford Classical Dictionary*, s.v. "Hippocrates."
3. Sargent, *Hippocratic Heritage*, esp. 46–62; Glacken, *Traces on the Rhodian Shore*, 82–88; Lloyd, *Hippocratic Writings*; Riley, *Eighteenth-Century Campaign*.
4. Sargent, *Hippocratic Heritage*, 63–80.
5. Bodin, *Methodus ad facilem historiarum cognitionem* (1566), quoted in Glacken, *Traces on the Rhodian Shore*, 439.
6. Burton, *Anatomy of Melancholy*, 1:237–41.
7. Burke, *Philosophical Enquiry*, 147.
8. Boyle, *General History of the Air*, 77.
9. Wren, *Parentalia* (1750), quoted in Dewhurst, *John Locke*, 18.
10. Sargent, *Hippocratic Heritage*, 129–60; Riley, *Eighteenth-Century Campaign*, 11–12.
11. [Jurin,] *Invitatio ad Observationes Meteorologicas*; Rusnock, *Correspondence of James Jurin*, 27–31.
12. Clifton, *Tabular Observations*; biographical information from the *Dictionary of National Biography*.
13. Arbuthnot, *Essay concerning the Effects of Air*, 141.
14. Rusnock, "Hippocrates, Bacon, and Medical Meteorology."
15. Sloane, *Voyage*, 1:xxxiii–xliii.
16. Wintringham, *Treatise of Endemic Diseases*, iii–vi.
17. Wintringham, *Commentarius Nosologicus*. Wintringham's journal was first published under this title in 1727; a second edition was prepared by his son (later Sir Clifton Wintringham) in 1733; and a third edition published (as part of the senior Wintringham's *Collected Works*) in 1752.
18. Pickering, "Scheme of a Diary of the Weather," 2.
19. Rusnock, "Hippocrates, Bacon, and Medical Meteorology," 144–45; Rusnock, *Vital Accounts*, 124–25.
20. Pickering, "Scheme of a Diary of the Weather," 14.
21. Huxham, *Observations on the Air and Epidemic Diseases*; Rusnock, *Correspondence of James Jurin*, 227, 321, 332–33, 524–26.
22. Hillary, *Account of the Principle Variations*.
23. Ibid., 14.
24. Ibid., 72.
25. Cheyne, *English Malady*, 199.
26. Rutty, *Essay towards a Natural History of the County of Dublin*, 2:272.
27. Rutty, *Chronological History*, xvi.
28. Fothergill, *Works*, 95.

29. [Bisset], *Essay on the Medical Constitution of Great Britain*, 6, 11.

30. [Smollett?], Review of Bisset, 187, 188.

31. [Bisset], *Candid and Satisfactory Answers*.

32. Thomas Short, "A Journal of the Weather at Sheffield from January 1727 to December 1755," Bodleian Library, Oxford, MS 35450; Short, *Comparative History of the Increase and Decrease of Mankind*, 119–53.

33. Weather Diary of Dr. John Bayly, Chichester, 1769–1773, National Meteorological Archive, 1:22.

34. Observations of Thomas Hughes, Physician, from 1 January 1771 to 17 April 1813, at Stroud, Gloucestershire, National Meteorological Archive.

35. [Addison et al.], *The Spectator*, 6:160 (no. 440, 25 July 1712).

36. Boyle, "A New Experiment," 5156.

37. Lister, "Discourse concerning the Rising and Falling of the Quicksilver," 792–93.

38. Smith, *Horological Disquisitions*, 77–78; Neve, *Baroscopologia*, 30.

39. Arbuthnot, *Essay concerning the Effects of Air*, 29, 39.

40. Short, *New Observations*, 420–23.

41. Phelps, *Human Barometer*, unnumbered pages in preface.

42. Ibid., 21.

43. More, *Enthusiasmus Triumphatus*, 12.

44. See especially Heyd, "Medical Discourse." Also relevant are Porter, "Rage of Party"; MacDonald, *Mystical Bedlam*, 1–12; and Vermeir, "'Physical Prophet.'"

45. Barker-Benfield, *Culture of Sensibility*, 1–153. On the general topic, see also Langford, *A Polite and Commercial People*, 461–518; Mullan, *Sentiment and Sociability*, 201–40; Vila, *Enlightenment and Pathology*; and Van Sant, *Eighteenth-Century Sensibility*, 1–15. On masculinity and effeminacy, see Carter, *Men and the Emergence of Polite Society*, 53–162.

46. On Cheyne, see Guerrini, *Obesity and Depression*; Barker-Benfield, *Culture of Sensibility*, 6–15; Porter, *Flesh in the Age of Reason*, 237–40; Porter, "Consumption"; and Porter, "Introduction."

47. Cheyne, *English Malady*, 48–60 (quotation on 56).

48. Ibid., 55.

49. Ibid., 58–59, 199. See also Porter, "Introduction," esp. xxvi–xxxii.

50. Falconer, *Remarks on the Influence of Climate*, 15, 48–50, 71, 123. See also Wilson, *Some Observations Relative to the Influence of Climate*, 284; Glacken, *Traces on the Rhodian Shore*, 601–5; and Gidal, "Civic Melancholy."

51. Robertson, *History of America*, 415.

52. Johnson, *The Idler*, no. 11 (24 June 1758), in Johnson, *Works*, 2:38.

53. Barker-Benfield, *Culture of Sensibility*, 26.

54. Porter, *Flesh in the Age of Reason*, 167–93.

55. Wintringham, *Treatise of Endemic Diseases*, 123.

56. Hillary, *Observations on the Changes of the Air*, ix–xi. Hillary's use of slavery as a metaphor for European settlers' submission to fashion is rather startling. The condition was, of course, a literal reality for African captives and their descendants throughout the West Indies.

57. Arbuthnot, *Essay concerning the Effects of Air*, 50, 107.

58. Barker-Benfield, *Culture of Sensibility*, 104–53.

59. On Short's statistical work, see Rusnock, *Vital Accounts*, 143–49; and Riley, *Eighteenth-Century Campaign*, 28–29.

60. Short, *Comparative History of the Increase and Decrease of Mankind*, 35–36, 57–58.

61. Rutty, *Essay towards a Natural History of the County of Dublin*, 7–8.

62. On these projects, see Riley, *Eighteenth-Century Campaign*; Corbin, *The Foul and the Fragrant*; and Hannaway, "Environment and Miasmata."

63. Schaffer, "Measuring Virtue"; Dolan, "Conservative Politicians."

64. Hales, *Vegetable Staticks*; Schofield, *Mechanism and Materialism*, 68–79; Stewart, *Rise of Public Science*, 231–34; Allen and Schofield, *Stephen Hales*.

65. Singer, "Sir John Pringle"; Golinski, *Science as Public Culture*, 106–10.

66. Schofield, *Enlightenment of Joseph Priestley*, 250–71; McEvoy, "Joseph Priestley"; Golinski, *Science as Public Culture*, 112–17.

67. Pringle, quoted in Schaffer, "Measuring Virtue," 285.

68. Schaffer, "Measuring Virtue"; Beretta, "Pneumatics vs. 'Aerial Medicine'"; Golinski, *Science as Public Culture*, 86, 117–19.

69. Golinski, *Science as Public Culture*, 119–28; Schaffer, "Measuring Virtue"; Beretta, "Pneumatics vs. 'Aerial Medicine.'"

70. Schofield, *Enlightened Joseph Priestley*, 93–119; Priestley, quoted in Golinski, *Science as Public Culture*, 78.

71. Golinski, *Science as Public Culture*, 110–12.

72. McEvoy, "Joseph Priestley"; Schofield, *Enlightened Joseph Priestley*; Anderson and Lawrence, *Science, Medicine and Dissent*.

73. Walker, *Philosophical Estimate*, 29–30. On Walker, see also the anonymous notice in *European Magazine and London Review* 21 (1792): 411–13.

74. Porter, *Doctor of Society*, 103, 106. See also Golinski, *Science as Public Culture*, 153–66.

75. Beddoes, quoted in Golinski, *Science as Public Culture*, 165.

76. *Anti-Jacobin Review*, quoted in Dolan, "Conservative Politicians," 38.

77. Beddoes, quoted in Golinski, *Science as Public Culture*, 175.

78. Peacock, *Nightmare Abbey / Crotchet Castle*, 92.

79. Johnson, *History of Rasselas*, 78, 81. See also the comments on the episode in Daston, "Enlightenment Fears," 119–20. It is perhaps symptomatic of the suspicion surrounding aerial improvement at the beginning of the nineteenth century that John Williams, introducing his remarks on climatic change, disavowed the identity of a "dogmatising Theorist" and the suggestion that he was making "impious" proposals. In defense, he reasserted the providentialist line that human efforts to control and enhance the quality of the air were consistent with divine design. (Williams, *Climate of Great Britain*, iii–iv, 1–2)

80. Hannaway, "Environment and Miasmata"; Pickstone, "Dearth, Dirt and Fever Epidemics"; Jordanova, "Earth Science and Environmental Medicine."

81. Chadwick, quoted in Hamlin, *Public Health and Social Justice*, 6.

82. Hamlin, "Environmental Sensibility in Edinburgh"; Hamlin, "Providence and Putrefaction."

CHAPTER SIX *Climate and Civilization*

1. Smith, "Language of Human Nature," 96.
2. On culture, see Williams, *Culture and Society*; on society, see Wagner, " 'An Entirely New Object of Consciousness.' "
3. For a survey of the general question, see Spitzer, "Milieu and Ambiance."
4. Du Bos, *Critical Reflections*, 2:186.
5. Arbuthnot, *Essay concerning the Effects of Air*, 148–49.
6. Ibid., 153–54.
7. Montesquieu, *Spirit of Laws*, 417–50.
8. Ibid., 245–46.
9. Ibid., 289.
10. For analysis of Montesquieu's treatment of the issue, see Wokler, "Anthropology and Conjectural History"; Carrithers, "Enlightenment Science of Society"; Carrithers, "Introduction," 44–51; and Romani, *National Character and Public Spirit*, 24–32.
11. Heffernan, "Historical Geographies of the Future."
12. Hume's "Of National Characters" appeared first in a volume of three essays published in 1748. He returned to the theme of physical and moral causes in "Of the Populousness of Ancient Nations," published in 1752. See Hume, *Essays: Moral, Political, and Literary*, xii–xiv, 197–215. When Hume wrote to Montesquieu in April 1749, he recorded that he had read *Spirit of the Laws* during the previous autumn. See Greig, *Letters of David Hume*, 1:133.
13. Hume, *Essays: Moral, Political, and Literary*, 207.
14. Ibid., 208n. On this footnote specifically, see Popkin, "Philosophical Basis of Eighteenth-Century Racism"; and Eze, "Hume, Race, and Human Nature."
15. Hume, *Essays: Moral, Political, and Literary*, 198.
16. Ibid., 204.
17. Berry, " 'Climate' in the Eighteenth Century," 287. See also Berry, *Social Theory of the Scottish Enlightenment*, 78–87.
18. Millar, *The Origin and Distinction of Ranks* (3rd ed., 1781), quoted in Wheeler, *Complexion of Race*, 183–84.
19. Ferguson, *Essay on the History of Civil Society* (1767), quoted in Wheeler, *Complexion of Race*, 187.
20. On the emergence of racial thinking, see Wheeler, *Complexion of Race*; Harrison, " 'The Tender Frame of Man' "; and Harrison, *Climates and Constitutions*.
21. Gerbi, *Dispute of the New World*, 177–79; Berry, *Social Theory of the Scottish Enlightenment*, 86. Gerbi provides a comprehensive account of the debate as to whether human and animal life had "degenerated" in the New World.
22. Robertson, *History of America*, 414–15.
23. Ibid., 259.
24. Ibid., 282–99 (quotation on 287). See also Gerbi, *Dispute of the New World*, 52–156 (on de Pauw), 165–69 (on Robertson).
25. Robertson, *History of America*, 257–59.
26. Ibid., 347, 418.

27. For a general discussion, see Smitten, "Impartiality in Robertson's *History of America*."

28. Dunbar, *Essays on the History of Mankind*, 221–49, 321–47; Berry, " 'Climate' in the Eighteenth Century."

29. Wood, *Aberdeen Enlightenment*, 134–38.

30. Wilson, *Some Observations Relative to the Influence of Climate*, 237–45.

31. Glacken, *Traces on the Rhodian Shore*, 601.

32. Falconer, *Remarks on the Influence of Climate*, 6, 21, 73.

33. This terminology, which derives from Durkheim rather than from Dunbar himself, is used in Berry, " 'Climate' in the Eighteenth Century."

34. Arnold, "Introduction."

35. Sloane, *Voyage*, ix, xxxi, xlvii.

36. Bisset, *Medical Essays and Observations*, 11–20.

37. Hillary, *Observations on the Changes of the Air*, i–xiii.

38. Fortune, *Franklin and His Friends*, 51–65.

39. Aldredge, *Weather Observers*, 204–18.

40. Lining, "Extracts of Two Letters," quotation on 493.

41. Chalmers, *Account of the Weather*, iii; Aldredge, *Weather Observers*, 219–22.

42. Chalmers, *Account of the Weather*, 10, 35.

43. Hillary, *Observations on the Changes of the Air*, ix–xi; Johnson, *Influence of Tropical Climates*, 459.

44. Currie, *Historical Account of the Climates and Diseases*, 112, 116.

45. Volney, *View of the Climate and Soil*, 282, 279.

46. James Johnson, in his *Influence of Tropical Climates*, 447–49, claimed that people were invoking the climate to excuse debauched behavior in tropical settlements. He insisted that surrender to sensual passions was voluntary and should be curtailed by religious and moral principles.

47. Harrison, *Climates and Constitutions*, 11–18, 45, 88.

48. Anderson, "Climates of Opinion."

49. See, for example, Mitchell, "Essay upon the Causes of Different Colours of People"; and Smith, *Essay on the Causes of the Variety of Complexion*.

50. Lind, *Essay on Diseases*.

51. Hunter, *Observations on the Diseases of the Army*. For background, see Kiple and Ornelas, "Race, War and Tropical Medicine."

52. Lind, *Essay on Diseases*, 51–52.

53. Hunter, *Observations on the Diseases of the Army*, 15.

54. Ibid., 36; Lind, *Essay on Diseases*, 151.

55. Harrison, *Climates and Constitutions*, 104.

56. Johnson, *Influence of Tropical Climates*, 1, 417.

57. Valenčius, *Health of the Country*, 22–34, 229–47.

58. Kupperman, "Puzzle of the American Climate"; Kupperman, "Fear of Hot Climates"; Meyer, *Americans and Their Weather*, 17–42.

59. Stearns, *Science in the British Colonies*, 188–89, 434, 448–50, 468; Cassedy, "Meteorology and Medicine."

60. Cassedy, "Meteorology and Medicine," 197–99; Stearns, *Science in the British Colonies*, 646–47; Eisenstadt, "Weather and Weather Forecasting," 195–213.

61. Volney, *View of the Climate and Soil*, 134.

62. Chalmers, *Account of the Weather*, 18. See also Kupperman, "Fear of Hot Climates."

63. Chalmers, *Account of Weather*, 47–56.

64. Eisenstadt, "Weather and Weather Forecasting," 41–55, 146–51; Hall, *Worlds of Wonder*, 79–80, 106–9, 221–22.

65. Hall, *Worlds of Wonder*, 108; Stearns, *Science in the British Colonies*, 648–52.

66. Delbourgo, "Common Sense"; Riskin, *Science in the Age of Sensibility*, 69–103.

67. Fleming, *Historical Perspectives on Climate Change*, 21–32.

68. Kalm, *Travels in North America*, 1:271, 275–77, 2:509, 513; Volney, *View of the Climate and Soil*, 266–78; Currie, *Historical Account of the Climates and Diseases*, 93.

69. Williamson, "Attempt to Account for the Change of Climate."

70. Robertson, *History of America*, 258.

71. Dunbar, *Essays on the History of Mankind*, 356.

72. Jefferson, *Notes on the State of Virginia*, 119; Miller, *Jefferson and Nature*, 41–42.

73. Currie, *Historical Account of the Climates and Diseases*, 403, 81–82.

74. Currie, *Historical Account of the Climates and Diseases*, 86. On the supposed changes in the European climate since ancient times, see Williamson, "Attempt to Account for the Change of Climate," 340–41; Feldman, "Ancient Climate in the Eighteenth and Nineteenth Century."

75. Rush, "Account of the Climate of Pennsylvania," 61. Volney noted that while Rush quibbled about the degree to which climate change could be measured, "these doubts vanish before the multitude of witnesses and positive facts" proving that the transformation had occurred. (Volney, *View of the Climate and Soil*, 269)

76. Ramsay, *Sketch of the Soil*, 8. See also Brunhouse, "David Ramsay."

77. Williams, *Natural and Civil History of Vermont*, 57.

78. Ibid., 57–65.

79. Belknap, *History of New-Hampshire*, 3:171–90.

80. Grove, *Green Imperialism*, 264–308.

81. Volney, *View of the Climate and Soil*, 269.

82. On the later history of this idea in America, see Valenčius, *Health of the Country*, 214–19.

83. Glacken, *Traces on the Rhodian Shore*, 587–91, 669, 680; Gerbi, *Dispute of the New World*, 3–34.

84. Robertson, *History of America*, 257–61 (quotations on 261), 270–72, 287–96.

85. Jefferson, *Notes on the State of Virginia*, 116.

86. Gerbi, *Dispute of the New World*, 252–68.

87. Miller, *Jefferson and Nature*, 63–76.

88. Ibid., 75. Samuel Stanhope Smith, whose *Essay on the Causes of the Variety of Complexion* was first published in the same year as Jefferson's work (1787), shared his view that Native Americans were more amenable to having their physical characteristics changed by civilization than were Africans. (Smith, *Essay on the Causes of the Variety of Complexion*, 45n)

89. Williams, *Natural and Civil History of Vermont*, 183, 158.

90. For other defenses of the natives by American authors, see Gerbi, *Dispute of the New World*, 240–52.
91. Volney, *View of the Climate and Soil*, 330–31.
92. Williams, *Natural and Civil History of Vermont*, vii.
93. Currie, *Historical Account of the Climates and Diseases*, 89n.
94. Indicative of the continuing close relationship between American and British thinking is the emergence in the early nineteenth century of the idea that the British climate was undergoing changes similar to those experienced on the other side of the Atlantic. John Williams developed this claim in a book published in 1806. He noted that "the Americans have already begun to experience an amelioration of climate by the introduction of the arts connected with Agriculture," and asserted that altered agricultural practices in Britain were bringing about increased atmospheric humidity, cooler summers, and milder winters. (*Climate of Great Britain*, 149)

CONCLUSION   *The Science of Weather*

1. They include Middleton, *History of the Barometer*; Middleton, *History of the Thermometer*; Frisinger, *History of Meteorology*; Shaw, *Meteorology in History*; and Wolf, *History of Science*, 1:274–341.
2. Boyle, *General History of the Air*. On Aristotelian meteorology, see Taub, *Ancient Meteorology*, 77–115; and Janković, *Reading the Skies*, 16–22.
3. 1703 Diary, 404.
4. Hooke, "Method for Making a History of the Weather," 175.
5. Kirwan, *Comparative View of Meteorological Observations*, 4.
6. Howard, *Climate of London*, 1:iii.
7. Howard, *Cycle of Eighteen Years*, 15.
8. Anderson, *Predicting the Weather*, 285.
9. Kuhn, "Mathematical versus Experimental Traditions."
10. Cavendish, "Account of the Meteorological Instruments Used."
11. Feldman, "Late Enlightenment Meteorology"; Kington, *Weather of the 1780s*, 3–17.
12. Hooke, "Method for Making a History of the Weather."
13. Janković, *Reading the Skies*, 103–24.
14. Daniell, *Meteorological Essays and Observations*, xiv.
15. Herschel, "Meteorology," 123.
16. Cannon, *Science in Culture*, 73–110; Morrell and Thackray, *Gentlemen of Science*, 517–23; Janković, "Ideological Crests versus Empirical Troughs" (quoting William Whewell and John Herschel on "fagging" on 25).
17. Classic works include Middleton, *History of the Barometer*; Middleton, *History of the Thermometer*; and Barnett, "Development of Thermometry." These can be compared with the sociologically informed work on instruments surveyed in Golinski, *Making Natural Knowledge*, 133–45.
18. Howard, *Climate of London*, 1:xxxvi.
19. Kirwan, *Estimate of the Temperature*, iii.
20. Howard, *Climate of London*, 1:iv.

21. Kirwan, *Comparative View of Meteorological Observations*, 4.

22. Ibid., 19−29.

23. Kirwan, *Estimate of the Temperature*, iv.

24. The geographical extent of atmospheric phenomena is discussed in Kirwan, *Of the Variations of the Atmosphere*, 120−33 (quotation on 122).

25. On these developments, see Anderson, *Predicting the Weather*, 171−233; Fleming, *Meteorology in America*; Friedman, *Appropriating the Weather*; and Monmonier, *Air Apparent*, 1−80.

26. Monmonier, *Air Apparent*, 236 n. 7.

27. Cook, *Edmond Halley*, 190−96.

28. Sorrenson, "The Ship as a Scientific Instrument."

29. On these projects, see Bedini, "Transit in the Tower"; Widmalm, "Accuracy, Rhetoric, and Technology"; and Alder, *Measure of All Things*. Also valuable is Licoppe, "Project for a Map of Languedoc."

30. In addition to the sources cited in the previous note, see also Bourguet and Licoppe, "Voyages, mesures et instruments"; and Bourguet, "Landscape with Numbers."

31. Livingstone, *Putting Science in Its Place*; Livingstone and Withers, *Geography and Enlightenment.*

32. Passy, "Tragedies of Nature."

33. Caldwell, "Was Katrina a Spiritual Message?"

34. Rothstein, "Seeking Justice."

{ BIBLIOGRAPHY }

MANUSCRIPTS

Bodleian Library, Oxford

John Locke. "Register of the Air." MS 48120, ff. 466–531.

Samuel Say. "A Journal of the Weather at Lostaff [Lowestoft] in Suffolk, from 1695 to 1724, by the Rev. Mr. Say." MS 35448.

Thomas Short. "A Journal of the Weather at Sheffield from January 1727 to December 1755." MS 35450.

British Library, London

Roger North. "Essay of the Barometer." Additional MSS 32,541. Two drafts, 297 ff.

———. "The Air-Gager. Being a Meteorologicall Essay of the Barometer." Additional MSS 32,542. Another draft of "Essay of the Barometer," 114 ff.

———. "An Essay concerning the Reason and Use of ye Baroscope." Additional MSS 32,453. 33 ff.

Gilbert White. "The Naturalist's Journal." Additional MSS 31,846–31,851. 6 vols., 1768–93.

Hereford and Worcestershire Record Office, Worcester

Hanbury Parish Register and Will of Thomas Appletree. Microfilm.

Houghton Library, Harvard University, Cambridge, MA

Letters by Samuel and Thomas Barker to Gilbert White. bMS Eng. 731(4). 13 February 1776.

Thomas Barker. "Notebook by Thomas Barker." bMS Eng. 737. 1730–1801.

Lancing College Archive, Lancing, West Sussex

1703 Weather Diary.

National Meteorological Archive, Bracknell, Berkshire

Weather Diary of Dr. John Bayly, Chichester, 1769–1773. 3 vols.

Observations of Thomas Hughes, Physician, from 1 January 1771 to 17 April 1813, at Stroud, Gloucestershire. 3 vols. With notes by C. E. Britton.

Observations by Joseph Tucker, 1730–1733, at Rye, Sussex.

P. R. Zealley. MS Copy of 18th Century Meteorological Diary Written by J. Whiston [*sic*], Worcestershire, 1703.

Royal Society, London
Classified Papers, vol. 5: Journals of the Weather, 1696–1725.
Testimonial by John Patrick, 13 April 1727. Early Letters, H3:127.

West Sussex Record Office, Chichester, West Sussex
1703 Weather Diary. Accession no. 12,761 (now returned to Lancing College).

Whipple Museum of the History of Science, Cambridge University
Peter Rabalio. "Rules for Foretelling the Alteration of the Weather by the Barometer."
Broadsheet, printed by J. Grundy, Worcester.

PRIMARY SOURCES (FIRST PUBLISHED BEFORE 1900)

Adams, George. *A Short Dissertation on the Barometer, Thermometer, and Other Meteo-rological Instruments.* London: R. Hindmarsh, 1790.
[Addison, Joseph, Richard Steele, et al.] *The Spectator.* Edited by G. Gregory Smith. 8 vols. in 4. London: J. M. Dent and Co., n.d.
[Aikin, John.] *The Calendar of Nature, Designed for the Instruction and Entertainment of Young Persons.* Warrington: W. Eyres for J. Johnson, 1784.
The Amazing Tempest. London: P. Mead, 1703.
Andrews, Henry. *A Royal Almanack and Meteorological Diary for the Year of Our Lord 1778.* London: for T. Carnan, 1778.
Arbuthnot, John. *An Essay concerning the Effects of Air on Human Bodies.* London: J. Tonson, 1733.
Aristotle. *Meteorologica.* Translated by H. D. P. Lee. Cambridge, MA: Harvard University Press, 1987.
The Art of Complaisance, or the Means to Oblige in Conversation. 2nd ed. London: John Starkey, 1677.
Barker, Thomas. "An Account of an Extraordinary Meteor Seen in the County of Rutland Which Resembled a Waterspout." *Philosophical Transactions* 46 (1749–50): 248–49.
———. *An Account of the Discoveries concerning Comets, with the Way to Find Their Orbits.* London: J. Whiston and B. White, 1757.
———. "Remarks on the Mutations of Stars." *Philosophical Transactions* 51 (1759–60): 498–504.
———. "An Account of a Remarkable Halo in a Letter to the Rev. William Stukeley." *Philosophical Transactions* 52 (1761–62): 3–6.
———. "Abstract of a Register of the Barometer, Thermometer, and Rain at Lyndon in Rutland, 1783." *Philosophical Transactions* 74 (1784): 283–86.
[Barlow, Edward.] *Meteorological Essays concerning the Origin of Springs, Generation of Rain, and Production of Wind.* London: John Hooke and Thomas Caldecott, 1715.
Bateman, Thomas. *Reports on the Diseases of London and the State of the Weather, from 1804 to 1816.* London: Longman, Hurst, Rees, Orme, and Brown, 1819.
Beddoes, Thomas. *Notice of Some Observations Made at the Medical Pneumatic Institution.* Bristol: Biggs and Cottle, 1799.

Belknap, Jeremy. *The History of New-Hampshire, Containing a Geographical Description of the State, with Sketches of Its Natural History*. 3 vols. Dover, NH: for O. Crosby and J. Varney by J. Mann and J. K. Remick, 1812.

Bent, William. *Eight Meteorological Journals of the Years 1793 to 1800, Kept in London… to Which Are Added Observations on the Diseases in the City and Its Vicinity*. London: for W. Bent, 1801.

Birch, Thomas. *The History of the Royal Society of London for Improving of Natural Knowledge*. 4 vols. London: A. Millar, 1756–57.

[Bisset, Charles.] *An Essay on the Medical Constitution of Great Britain*. London: A. Millar and D. Wilson, 1762.

[———.] *Candid and Satisfactory Answers to the Several Criticisms of the Critical Reviewers on an Essay on the Medical Constitution of Great Britain*. London: A. Millar and D. Wilson, 1763.

———. *Medical Essays and Observations*. Newcastle-upon-Tyne: I. Thompson, 1766.

Bohun, Ralph. *A Discourse concerning the Origine and Properties of Wind*. Oxford: W. Hall for Thomas Bowman, 1671.

Boswell, James. *Life of Johnson*. London: Oxford University Press, 1953.

Boyle, Robert. "Some Observations and Directions about the Barometer." *Philosophical Transactions* 1, no.11 (1665–66): 181–84.

———. "A New Experiment . . . concerning an Effect of the Varying Weight of the Atmosphere upon Some Bodies in the Water." *Philosophical Transactions* 7, no. 91 (1672–73): 5156–59.

———. *The General History of the Air, Designed and Begun by the Honble. Robert Boyle, Esq*. London: Awnsham and John Churchill, 1692.

———. *The Works of the Honourable Robert Boyle*. Edited by Thomas Birch. 2nd ed. 6 vols. London: J. and F. Rivington, 1772.

Bradbury, Thomas. *God's Empire over the Wind, Consider'd in a Sermon on the Fast-Day, January 19, 1703/4*. London: S. Bridge for Jonathan Robinson, 1704.

Budgen, Richard. *The Passage of the Hurricane from the Sea-side at Bexhill in Sussex to Newingden-Level*. London: John Senex, 1730.

Burke, Edmund. *A Philosophical Enquiry into the Origin of Our Ideas of the Sublime and the Beautiful*. Edited by James T. Boulton. Notre Dame, IN: University of Notre Dame Press, 1968.

Burton, Robert. *The Anatomy of Melancholy*. Edited by Holbrook Jackson. 3 vols. London: J. M. Dent and Sons, 1932.

Campbell, John. *A Political Survey of Britain*. 2 vols. London: Richardson and Urquhart et al., 1774.

Cavendish, Henry. "An Account of the Meteorological Instruments Used at the Royal Society's House." *Philosophical Transactions* 66 (1776): 375–401.

Chalmers, Lionel. *An Account of the Weather and Diseases of South Carolina*. 2 vols. in 1. London: Edward and Charles Dilley, 1776.

Cheyne, George. *The English Malady, or a Treatise of Nervous Diseases of All Kinds, as Spleen, Vapours, Lowness of Spirits, Hypochondriacal, and Hysterical Distempers, &c*. London: G. Strahan, 1733.

————. *The English Malady* [1733]. Edited by Roy Porter. London: Tavistock/Routledge, 1991.

Claridge, John. *The Shepheards' Legacy, or John Claridge His Forty Years Experience of the Weather*. London: John Hancock, 1670.

————. *The Shepherd of Banbury's Rules to Judge of the Changes of the Weather*. [Edited by John Campbell.] London: W. Bickerton, 1744.

Clifton, Francis. *Tabular Observations Recommended as the Plainest and Surest Way of Practising and Improving Physick*. London: J. Brindley, 1731.

Cock, William. *Meteorologia, or the True Way of Foreseeing and Judging the Inclination of the Air and the Alteration of the Weather in Several Regions*. London: John Conyers, 1671.

A Companion to the Weather-Glass, or the Nature, Construction, and Use of the Barometer, Thermometer, and Hygrometer. Edinburgh: T. Ross for J. Guthrie, 1796.

[Constable, John.] *The Conversation of Gentlemen Considered, in Most of the Ways That Make Their Mutual Company Agreeable, or Disagreeable*. London: J. Hoyles, 1738.

Cullum, John. "An Account of a Remarkable Frost on the 23rd of June 1783." *Philosophical Transactions* 74 (1784): 416–18.

Currie, William. *An Historical Account of the Climates and Diseases of the United States of America, and of the Remedies and Methods of Treatment Which Have Been Found Most Useful and Efficacious*. Philadelphia: T. Dobson, 1792.

[D'Ancourt.] *The Lady's Preceptor, or a Letter to a Young Lady of Distinction upon Politeness. Taken from the French . . . by a Gentleman of Cambridge*. London: J. Watts, 1743.

Daniell, John Frederic. *Meteorological Essays and Observations*. London: Thomas and George Underwood, 1823.

[Defoe, Daniel.] *The Storm, or a Collection of the Most Remarkable Casualties and Disasters Which Happen'd in the Late Dreadful Tempest*. London: G. Sawbridge, 1704.

[————.] *An Elegy on the Author of the True-Born-English-Man, with an Essay on the Late Storm*. London, 1708.

————. *The Storm*. Edited by Richard Hamblyn. London: Penguin Books, 2005.

Denham, M. A. *A Collection of Proverbs and Popular Sayings Relating to the Seasons, the Weather, and Agricultural Pursuits*. London: Percy Society, 1846.

Derham, William. "A Letter for the Reverend Mr. *William Derham*, F.R.S., Containing His Observations concerning the Late Storm." *Philosophical Transactions* 24, no. 289 (1704–5): 1530–34.

————. *Physico-Theology, or a Demonstration of the Being and Attributes of God from His Works of Creation*. London: W. Innys, 1713.

Desaguliers, John Theophilus. *Lectures of Experimental Philosophy*. 2nd ed. London: W. Mears, B. Creake, and J. Sackfield, 1719.

A Diary or Weather-Journall Kept at [blank] in the Month of [blank] Anno Dom: [blank]. London: John Warner, [c. 1685?]

Du Bos, Jean-Baptiste, abbé. *Critical Reflections on Poetry and Painting*. Translated by Thomas Nugent from the 5th edition. 3 vols. London: John Nourse, 1748.

Dunbar, James. *Essays on the History of Mankind in Rude and Cultivated Ages*. 2nd ed. London: W. Strahan, T. Cadell, J. Balfour, 1781.

An Essay concerning the Late Apparition in the Heavens on the Sixth of March, Proving… *That It… Must of Necessity Be a Prodigy.* London: J. Morphew, 1716.

An Exact Relation of the Late Dreadful Tempest, or a Faithful Account of the Most Re- *markable Disasters Which Happened on That Occasion.* London: A. Baldwin, 1704.

Falconer, William. *Remarks on the Influence of Climate, Situation, Nature of Country,* *Population, Nature of Food, and Way of Life on the Disposition and Temper, Manners* *and Behaviour, Intellects, Customs, Form of Government, and Religion of Mankind.* London: C. Dilly, 1781.

[Forrester, James.] *The Polite Philosopher, or an Essay on That Art Which Makes a Man* *Happy in Himself, and Agreeable to Others.* Edinburgh: Robert Freebairn, 1734.

Forster, Thomas. *The Perennial Calendar and Companion to the Almanack.* London: Harding, Mavor, and Lepard, 1824.

Foster, John. *Alumni Oxonienses: The Members of the University of Oxford, 1500–1714.* 4 vols. Oxford: Parker and Co., 1891.

Fothergill, John. *The Works of John Fothergill, M.D.* Edited by John Coakley Lettsom. London: Charles Dilly, 1784.

Franklin, Benjamin. "Meteorological Imaginations and Conjectures." *Memoirs of the* *Literary and Philosophical Society of Manchester* 2 (1785): 357–61.

Fuller, John. "Part of a Letter from *John Fuller* of *Sussex*, Esq., concerning a Strange Effect of the Late Great Storm in That County." *Philosophical Transactions* 24, no. 289 (1704–5): 1530.

Fuller, Thomas. *Gnomologia: Adagies and Proverbs, Wise Sentences and Witty Say-* *ings, Ancient and Modern, Foreign and British.* London: B. Barker, A. Bettesworth, C. Hitch, 1732.

[Gadbury, John.] *Stars and Planets the Best Barometers and Truest Interpreters of All* *Airy Vicissitudes.* London: John Nutt, 1701.

Garden, George. "A Discourse concerning Weather." *Philosophical Transactions* 15, no. 171 (1685): 991–1001.

Gifford, A. *A Sermon in Commemoration of the Great Storm.* London: Aaron Ward, 1733.

Goad, John. *Astro-Meteorologica, or Aphorisms and Discourses of the Bodies Cœlestial,* *Their Natures and Influences.* London: J. Rawlins, 1686.

[Hale, Matthew.] *Difficiles Nugae, or Observations Touching the Torricellian Experiment* *and the Various Solutions of the Same.* London: W. Godbid for William Shrowsbury, 1674.

Hales, Stephen. *Vegetable Staticks.* London: Scientific Book Guild, 1961.

Hall, Allen. *Observations on the Weather.* 5th ed. Lincoln: Drury, 1788.

Halley, Edmond. "A Discourse of the Rule of the Decrease of the Height of the Mercury in the Barometer … with an Attempt to Discover the True Reason of the Rising and Falling of the Mercury upon Change of Weather." *Philosophical Transactions* 16, no. 181 (1686): 104–16.

———. "An Historical Account of the Trade Winds and Monsoons." *Philosophical* *Transactions* 16, no. 183 (1686): 153–68.

Hamilton, William. "An Account of the Earthquakes Which Happened in Italy from February to May 1783." *Philosophical Transactions* 73 (1783): 169–208.

Harris, John. *Lexicon Technicum, or an Universal Dictionary of Arts and Sciences*. London: Daniel Brown et al., 1704.

Herschel, John F. W. "Meteorology." In *A Manual of Scientific Enquiry, Prepared for the Use of Officers in Her Majesty's Navy and Travellers in General*, edited by Robert Main, 120–69, 3rd ed. London: John Murray, 1859. Originally published 1849.

———. *Meteorology: From the Encyclopaedia Britannica*. Edinburgh: Adam and Charles Black, 1861.

———. "The Weather and Weather Prophets." In Herschel, *Familiar Lectures on Scientific Subjects*, 142–75. London: Alexander Strahan, 1866.

Hillary, William. *A Practical Essay on the Small-Pox*. 2nd ed. London: C. Hitch, 1740.

———. *An Account of the Principle Variations of the Weather and the Concomitant Epidemical Diseases, from the Year 1726 to the End of the Year 1734, as They Appeared at Ripon and the Adjacent Parts of the County of York*. Bound with the above (separately paginated).

———. *Observations on the Changes of the Air and the Concomitant Epidemic Diseases in the Island of Barbadoes*. 2nd ed. London: L. Hawes, W. Clarke, R. Collins, 1766.

Hooke, Robert. "Method for Making a History of the Weather." In Sprat, *History of the Royal Society*, 173–79.

Howard, Luke. *The Climate of London Deduced from Meteorological Observations Made at Different Places in the Neighbourhood of the Metropolis*. 2 vols. London: W. Phillips, 1818–20.

———. *Seven Lectures on Meteorology*. Pontefract: James Lucas, 1837.

———. *A Cycle of Eighteen Years in the Seasons of Britain*. London: James Ridgway, 1842.

Hume, David. *Essays: Moral, Political, and Literary*. Edited by Eugene F. Miller. Indianapolis, IN: Liberty Classics, 1985.

Hunter, John. *Observations on the Diseases of the Army in Jamaica and on the Best Means of Preserving the Health of Europeans in that Climate*. London: for G. Nichol, 1788.

Hussey, Joseph. *A Warning from the Winds: A Sermon Preach'd upon Wednesday, January XIX, 1703/4*. London: William and Joseph Marshall, 1704.

Hutchinson, Benjamin. *A Calendar of the Weather for the Year 1781, with an Introductory Discourse on the Moon's Influence at Common Lunations in General*. London: J. Fielding, 1782.

Huxham, John. *Observations on the Air and Epidemic Diseases from the Year MDCCXXVIII to MDCCXXXVII Inclusive*. London: J. Hinton, 1759.

Inwards, Richard. *Weather Lore: A Collection of Proverbs, Sayings, and Rules concerning the Weather*. 3rd ed. London: Elliot Stock, 1898.

Jefferson, Thomas. *Notes on the State of Virginia* [1787]. In *The Portable Thomas Jefferson*, edited by Merrill D. Peterson, 23–232. New York: Viking Press, 1975.

Johnson, James. *The Influence of Tropical Climates, More Especially the Climate of India, on European Constitutions*. London: J. J. Stockdale, 1813.

Johnson, Samuel. *The History of Rasselas, Prince of Abyssinia*. In *Shorter Novels: Eighteenth Century*, edited by Philip Henderson, 1–95. London: Dent, 1930.

[Jurin, James.] *Invitatio ad Observationes Meteorologicas Communi Consilio Instituendas*. London: William and John Innys, 1724.

Kalm, Pehr. *Peter Kalm's Travels in North America.* Edited by Adolph B. Benson, 2 vols. New York: Wilson-Erickson, 1937.

Kirwan, Richard. *An Estimate of the Temperature of Different Latitudes.* London: J. Davis, 1787.

———. "An Essay on the Variations of the Barometer." *Transactions of the Royal Irish Academy* 2 (1788): 43–73.

———. *A Comparative View of Meteorological Observations Made in Ireland since the Year M.DCC.LXXXVIII.* Dublin: George Bonham, 1794.

———. *Of the Variations of the Atmosphere.* Dublin: George Bonham, 1801.

The Knowledge of Things Unknown. London: for J. Clarke and A. Wilde, 1743.

Laurence, John. *The Fruit-Garden Kalendar . . . to Which Is Added an Appendix of the Usefulness of the Barometer.* London: Bernard Lintot, 1718.

The Lady's Preceptor, or a Letter to a Young Lady of Distinction upon Politeness. London: J. Watts, 1743.

Leeuwenhoek, Anton van. "Part of a Letter from Mr. *Anthony van Leeuwenhoek,* F.R.S., Giving His Observations on the Late Storm." *Philosophical Transactions* 24, no. 289 (1704–5): 1535–37.

Leigh, Charles. *The Natural History of Lancashire, Cheshire, and the Peak in Derbyshire.* Oxford: for the author, 1700.

Lind, James. *An Essay on Diseases Incidental to Europeans in Hot Climates, with the Method of Preventing their Fatal Consequences.* London: T. Becket and P. A. De Hondt, 1768.

Lining, John. "Extracts of Two Letters from Dr. John Lining . . . Giving an Account of Statical Experiments . . . Accompanied with Meteorological Observations ..." *Philosophical Transactions* 42, no. 470 (1742–43): 491–509.

Lister, Martin. "Three Papers of Dr. Martin Lyster, the First of the Nature of Earthquakes." *Philosophical Transactions* 14, no. 157 (1684): 512–17.

———. "A Discourse concerning the Rising and Falling of the Quicksilver in the Barometer." *Philosophical Transactions* 14, no. 165 (1684): 790–94.

Magellan, Jean Hyacinthe de. *Description et usage des nouveaux baromètres.* London: W. Richardson, 1779.

Marshall, William. *Experiments and Observations concerning Agriculture and the Weather.* London: J. Dodsley, 1779.

Martin, Benjamin. *A Description of the Nature, Construction, and Use of the Torricellian or Simple Barometer.* London: for the author, 1766.

———. *The Young Gentleman and Lady's Philosophy, in a Continued Survey of the Works of Nature and Art, by Way of Dialogue.* 3rd ed. 3 vols. London: W. Owen, 1781.

The Meteorologist's Assistant in Keeping a Diary of the Weather. London: for the author, 1793.

Miller, Philip. *The Gardener's Dictionary.* 2 vols. London: J. and J. Rivington, 1752.

Mills, John. *An Essay on the Weather, with Remarks on the Shepherd of Banbury's Rules for Judging of Its Changes.* London: S. Hooper, 1770.

Mitchell, John. "An Essay upon the Causes of Different Colours of People in Different Climates." *Philosophical Transactions* 43, no. 474 (1744–45): 102–50.

Molesworth, Caroline. *The Cobham Journals: Abstracts and Summaries of Meteorological and Phenological Observations Made by Miss Caroline Molesworth.* Edited by Eleanor A. Ormerod. London: Edward Stanford, 1880.

Montesquieu, Charles Louis de Secondat, Baron de. *The Spirit of Laws.* A compendium of the first English edition (1750), with an English translation of *An Essay on Causes Affecting Minds and Characters.* Edited by David Wallace Carrithers. Berkeley and Los Angeles: University of California Press, 1977.

Moore, Francis. *Vox Stellarum, or a Loyal Almanack for the Year of Human Redemption M,DCC,XCI.* London: for the Company of Stationers, 1791.

More, Henry. *Enthusiasmus Triumphatus, or a Brief Discourse of the Nature, Causes, Kinds, and Cure of Enthusiasm.* London: James Flesher for William Morden, 1662.

Neve, Richard. *Baroscopologia, or a Discourse of the Baroscope or Quicksilver Weather-Glass.* London: W. Keble, 1708.

[Oldenburg, Henry, and John Beale.] "A Relation of Some Mercurial Observations and Their Results." *Philosophical Transactions* 1, no. 9 (1666): 153–59.

———. "Observations Continued upon the Barometer, or Rather Ballance of the Air." *Philosophical Transactions* 1, no. 10 (1666): 163–66.

Parker, Gustavus. *An Account of a Portable Barometer, with Reasons and Rules for the Use of It.* London: William Haws, 1700.

[———.] *A New Account of the Alterations of Wind and Weather by the Discoveries of the Portable Barometer.* London: W. Hawes, 1700.

[———.] *A New Baroscopical Account of the Daily Alterations of the Wind and Weather.* London: W. Hawes, 1701.

[Parker, Gustavus, and John Patrick.] *A New Account of the Alterations of the Wind and Weather, by the Discoveries of the Portable Barometer … [with] A Journal of the Wind and Weather … as It Was Observ'd by J. Patrick, Barometer Maker.* London: J. Nutt, [1700].

Partridge, John. *Merlinus Redivivus: Being an Almanack for the Year of Our Redemption, 1685.* London: R. R. for the Company of Stationers, [1685].

Patrick, John. *A New Improvement of the Quicksilver Barometer, Made by John Patrick.* London: Richard Newcomb, [1710?]

Peacock, Thomas Love. *Nightmare Abbey / Crotchet Castle.* Edited by Raymond Wright. London: Penguin Books, 1969.

Phelps, John. *The Human Barometer, or Living Weather-Glass.* London: M. Cooper, 1743.

Pickering, Roger. "A Scheme of a Diary of the Weather; Together with Draughts and Descriptions of Machines Subservient Thereunto." *Philosophical Transactions* 43 (1744–45): 1–18.

Plot, Robert. *The Natural History of Oxfordshire, Being an Essay toward the Natural History of England.* Oxford: at the Theater, 1677.

———. "A Letter … to Dr. Martin Lister, F. of the R. S., concerning the Use Which May Be Made of the Following History of the Weather." *Philosophical Transactions* 15, no. 169 (1684): 930–43.

———. *The Natural History of Staffordshire.* Oxford: at the Theater, 1686.

Pointer, John. *A Rational Account of the Weather, Shewing the Signs of Its Several Changes and Alterations.* Oxford: L. L. for S. Wilmot, 1723.

A Practical Discourse on the Late Earthquakes: with an Historical Account of Prodigies and Their Various Effects. London: J. Dunton, 1692.

Priestley, Joseph. *Directions for Impregnating Water with Fixed Air.* London: J. Johnson, 1772.

———. *Experiments and Observations on Different Kinds of Air.* 2nd ed. 3 vols. London: J. Johnson, 1775–77.

Pringle, John. *Observations on the Diseases of the Army in Camp and Garrison.* London: A. Millar et al., 1752.

Ramsay, David. *A Sketch of the Soil, Climate, Weather, and Diseases of South Carolina.* Charleston, SC: W. P. Young, 1796.

Ray, John. *A Collection of English Proverbs, Digested into a Convenient Method.* 2nd ed. Cambridge: John Hayes, 1678.

[Richardson, Samuel.] *Clarissa: Or the History of a Young Lady, Comprehending the Most Important Concerns of Private Life.* 7 vols. London: for S. Richardson, 1748.

Robertson, William. *The History of America.* London: W Strahan, T. Cadell, and J. Balfour, 1777.

Robinson, Thomas. *New Observations on the Natural History of This World of Matter and This World of Life.* London: John Newton, 1696.

[Rooke, Hayman.] *A Meteorological Register, Kept at Mansfield Woodhouse in Nottinghamshire.* Nottingham: S. Tupman, 1795.

Roubais du Tourcoin, Jacques de. *A Physical Dissertation concerning the Cause of the Variation of the Barometer.* London: J. Peele, 1721.

Rush, Benjamin. "An Account of the Climate of Pennsylvania and Its Influence upon the Human Body." In Rush, *Medical Inquiries and Observations,* 57–88. Philadelphia: Prichard and Hall, 1789.

Rutty, John. *A Chronological History of the Weather and Seasons and of the Prevailing Diseases in Dublin.* London: Robinson and Roberts, 1770.

———. *An Essay towards a Natural History of the County of Dublin, Accommodated to the Noble Designs of the Dublin Society.* 2 vols. Dublin: W. Sleater, 1772.

———. *A Spiritual Diary and Soliloquies.* 2 vols. London: James Phillips, 1776.

"S. W." *A Poem on the Late Violent Storm.* London: B. Bragg, 1703.

Saul, Edward. *An Historical and Philosophical Account of the Barometer or Weather-Glass.* London: A. Betteswood, 1730.

Seneca. *Natural Questions.* Translated by Thomas H. Corcoran. 2 vols. Cambridge, MA: Harvard University Press, 1971.

[Short, Thomas.] *A General Chronological History of the Air, Weather, Seasons, Meteors, &c.* 2 vols. London: T. Longman and A. Millar, 1749.

———. *New Observations, Natural, Moral, Civil, Political, and Medical on City, Town, and Country Bills of Mortality.* London: T. Longman and A. Millar, 1750.

———. *A Comparative History of the Increase and Decrease of Mankind in England and Several Countries Abroad, according to the Different Soils, Situations, Business of Life, Use of the Non-Naturals, &c.* London: for W. Nicholl, 1767.

Sinclair, George. *The Principles of Astronomy and Navigation . . . to Which Is Added, A Discovery of the Secrets of Nature Which Are Found in the Mercurial Weather-Glass.* Edinburgh: the heir of Andrew Anderson, 1688.

Sloane, Hans. *A Voyage to the Islands Madera, Barbados, Nieves, S. Christophers and Jamaica, with the Natural History . . . of the Last of Those Islands.* 2 vols. London: B. M. for the author, 1707, 1725.

Smith, John. *A Compleat Discourse of the Nature, Use, and Right Managing of That Wonderful Instrument the Baroscope or Quick-Silver Weather-Glass.* London: Joseph Watts, 1688.

————. *Horological Disquisitions concerning the Nature of Time . . . to Which Is Added, The Best Rules for the Ordering and Use Both of the Quick-Silver and Spirit Weather-Glasses.* London: Richard Cumberland, 1694.

Smith, Samuel Stanhope. *An Essay on the Causes of the Variety of Complexion and Figure in the Human Species.* Reprinted from the 2nd edition, 1810. Edited by Winthrop D. Jordan. Cambridge, MA: Belknap Press of Harvard University Press, 1965.

[Smollett, Tobias?] Review of Bisset, *Essay on the Medical Constitution of Great Britain. The Critical Review or Annals of Literature, by a Society of Gentlemen,* 14 (1763; number for September 1762): 186–89.

[————.] *The Adventures of Peregrine Pickle, in Which Are Included Memoirs of a Lady of Quality.* London: Oxford University Press, 1964.

Sprat, Thomas. *The History of the Royal Society of London.* London: T. R. for J. Martyn, 1667.

Sterne, Laurence. *The Life and Opinions of Tristram Shandy, Gentleman.* Edited by Graham Petrie. Harmondsworth, Middlesex: Penguin Books, 1967.

Stillingfleet, Benjamin. *The Calendar of Flora, Swedish and English, Made in the Year 1755.* London: for the author, 1761.

Suetonius, Gaius Tranquillus. *The Twelve Caesars.* Translated by Robert Graves. Harmondsworth, Middlesex: Penguin Books, 1969.

Swainson, Charles. *A Handbook of Weather Folk-Lore.* Edinburgh: William Blackwood, 1873.

Swift, Jonathan. *Swift's Polite Conversation.* Edited by Eric Partridge. London: Andre Deutsch, 1963.

————. *Gulliver's Travels.* Edited by Peter Dixon and John Chalker. Harmondsworth, Middlesex: Penguin Books, 1967.

Taylor, Joseph. *The Complete Weather Guide: A Collection of Practical Observations for Prognosticating the Weather Drawn from Plants, Animals, Inanimate Bodies, and Also by Means of Philosophical Instruments.* London: John Harding, 1812.

The Terrible Stormy Wind and Tempest. London: W. Freeman, 1705.

"To the Authors of the *Universal Magazine.*" *Universal Magazine of Knowledge and Pleasure,* no. 1 (June 1747): 18–20; no. 2 (July 1747): 57–59.

Townley, Richard. *A Journal Kept in the Isle of Man, Giving an Account of the Wind and Weather and Daily Occurrences for Upwards of Eleven Months.* 2 vols. Whitehaven: J. Ware and Son, 1791.

[Trenchard, John.] *The Natural History of Superstition.* London: A. Baldwin, 1709.

A True and Particular Account of a Storm of Thunder & Lightning. London: John Morphew, 1711.

Volney, Constantin François de. *View of the Climate and Soil of the United States of America, to Which Are Annexed Some Accounts of Florida . . . and the Savages or Natives.* English translation. London: for J. Johnson, 1804.

"W. F." [William Fulke]. *Meteors, or a Plain Description of All Kinds of Meteors, as Well Fiery and Ayrie, as Watry and Earthy.* London: William Leake, 1670.

Walker, Adam. *A Philosophical Estimate of the Causes, Effects, and Cure of Unwholesome Air in Large Cities.* London: for the author, 1778.

Wallis, John. "A Discourse concerning the Air's Gravity." *Philosophical Transactions* 15, no. 171 (1685): 1002–14.

Whewell, William. *Astronomy and General Physics Considered with Reference to Natural Theology.* 6th ed. London: William Pickering, 1847.

Whiston, William. *An Account of a Surprizing Meteor Seen in the Air, March the 6th, 1715/16 at Night.* 2nd ed. London: J. Senex, 1716.

White, Gilbert. *The Natural History and Antiquities of Selborne.* New ed. London: White, Cochrane, and Co., 1813.

Williams, John. *The Climate of Great Britain, or Remarks on the Changes It Has Undergone, Particularly within the Last Fifty Years.* London: for C. and R. Baldwin, 1806.

Williams, Samuel. *The Natural and Civil History of Vermont.* Walpole, NH: Isaiah Thomas and David Carlisle Jr., 1794.

Williamson, Hugh. "An Attempt to Account for the Change of Climate Which Has Been Observed in the Middle Colonies in North-America." *Transactions of the American Philosophical Society*, 2nd ed., 1 (1789): 336–45.

Willsford, Thomas. *Nature's Secrets, or the Admirable and Wonderful History of the Generation of Meteors and Blazing Stars.* London: N. Brooke, 1665.

Wilson, Alexander. *Some Observations relative to the Influence of Climate on Vegetable and Animal Bodies.* London: T. Cadell, 1780.

Wintringham, Clifton. *A Treatise of Endemic Diseases, Wherein the Different Nature of Airs, Situations, Soils, Waters, Diet, &c. Are Mechanically Explain'd and Accounted for.* York: Grace White, 1718.

———. *Commentarius Nosologicus: A Treatise in the Study of Diseases.* Translated by Esme Johnson from the 3rd edition, 1752. Pocklington: Joint Committee of the Class of the Workers' Educational Association and the University of Hull, 1979.

A Wonderful History of All the Storms, Hurricanes, Earthquakes, &c. London: A. Baldwin, 1704.

Wood, Anthony à. *Athenae Oxonienses: An Exact History of All the Writers and Bishops Who Have Had Their Education in the University of Oxford.* Edited by Philip Bliss. 3rd ed. 4 vols. London: F. C. and J. Rivington et al., 1813–20.

———. *Fasti Oxonienses, or Annals of the University of Oxford.* Edited by Philip Bliss. 3rd ed. London: F. C. and J. Rivington et al., 1815.

Wood, Loftus. *The Valetudinarian's Companion, or Observations on Air, Exercise, and Regimen, with the Medical Properties of the Sea and Mineral Waters of Brighthelmston.* London: C. Watts, 1782.

SECONDARY SOURCES (FIRST PUBLISHED AFTER 1900)

Albury, W. R. "Halley and the Barometer." In *Standing on the Shoulders of Giants: A Longer View of Newton and Halley*, edited by Norman J. W. Thrower, 220–27. Berkeley and Los Angeles: University of California Press, 1990.

Alder, Ken. *The Measure of All Things: The Seven-Year Odyssey That Transformed the World*. New York: Free Press, 2002.

Aldredge, Robert Croom. *Weather Observers and Observations at Charleston, South Carolina, 1670–1871*. Reprinted from *Historical Appendix of the Year Book of the City of Charleston for the Year 1940*, 190–257. Charleston, SC, n.d.

Allen, D. G. C., and Robert E. Schofield. *Stephen Hales: Scientist and Philanthropist*. London: Scolar Press, 1980.

Anderson, Katharine. "The Weather Prophets: Science and Reputation in Victorian Meteorology." *History of Science* 37 (1999): 179–216.

———. *Predicting the Weather: Victorians and the Science of Meteorology*. Chicago: University of Chicago Press, 2005.

Anderson, R. G. W., and Christopher Lawrence, eds. *Science, Medicine and Dissent: Joseph Priestley (1733–1804)*. London: Wellcome Trust / Science Museum, 1987.

Anderson, Warwick. "Climates of Opinion: Acclimatization in Nineteenth-Century France and England." *Victorian Studies* 35 (1991–92): 135–57.

Arnold, David. "Introduction: Tropical Medicine before Manson." In Arnold, *Warm Climates and Western Medicine*, 1–19.

———, ed. *Warm Climates and Western Medicine: The Emergence of Tropical Medicine, 1500–1900*. Amsterdam: Rodopi, 1996.

Arora, Shirley L. "Weather Proverbs: Some 'Folk' Views." *Proverbium* 8 (1991): 1–17.

Atkinson, A. D. "William Derham, F.R.S. (1657–1735)." *Annals of Science* 8 (1952): 368–92.

Baker, Keith Michael, and Peter Hanns Reill, eds. *What's Left of Enlightenment? A Postmodern Question*. Stanford, CA: Stanford University Press, 2001.

Baldick, Chris. *In Frankenstein's Shadow: Myth, Monstrosity, and Nineteenth-Century Writing*. Oxford: Clarendon Press, 1987.

Banfield, Edwin. *Antique Barometers: An Illustrated Survey*. Trowbridge, Wiltshire: Baros Books, 1976.

———. *Barometer Makers and Retailers, 1660–1900*. Trowbridge, Wiltshire: Baros Books, 1991.

Bann, Stephen, ed. *Frankenstein, Creation and Monstrosity*. London: Reaktion Books, 1994.

Barker-Benfield, G. J. *The Culture of Sensibility: Sex and Society in Eighteenth-Century Britain*. Chicago: University of Chicago Press, 1992.

Barnett, Martin K. "The Development of Thermometry and the Temperature Concept." *Osiris*, 1st ser., 12 (1956): 269–341.

Bate, Jonathan. "Living with the Weather." *Studies in Romanticism* 35 (1996): 431–47.

Bedini, Silvio A. "The Transit in the Tower: English Astronomical Instruments in Colonial America." *Annals of Science* 54 (1997): 161–96.

Beier, Lucinda McCray. *Sufferers and Healers: The Experience of Illness in Seventeenth-Century England.* London: Routledge and Kegan Paul, 1987.

Benstead, C. R. *The Weather Eye: An Irreverent Discourse upon Meteorological Lore, Ancient and Modern.* London: Robert Hale, [1940].

Beretta, Marco. "Pneumatics vs. 'Aerial Medicine': Salubrity and Respirability of Air at the End of the Eighteenth Century." In *Nuova Voltiana: Studies on Volta and His Times*, 2:49–71. Pavia: Università degli Studi di Pavia, 2000.

Berry, Christopher J. "'Climate' in the Eighteenth Century: James Dunbar and the Scottish Case." *Texas Studies in Literature and Language* 16 (1974): 281–92.

———. *Social Theory of the Scottish Enlightenment.* Edinburgh: Edinburgh University Press, 1997.

Bewell, Alan. *Romanticism and Colonial Disease.* Baltimore: Johns Hopkins University Press, 1999.

Blackburn, Bonnie, and Leofranc Holford-Strevens. *The Oxford Companion to the Year.* Oxford: Oxford University Press, 1999.

Boia, Lucian. *The Weather in the Imagination.* London: Reaktion Books, 2005.

Bourguet, Marie-Noëlle. "Landscape with Numbers: Natural History, Travel and Instruments in the Late Eighteenth and Early Nineteenth Centuries." In Bourguet, Licoppe, and Sibum, *Instruments, Travel, and Science*, 96–125.

Bourguet, Marie-Noëlle, and Christian Licoppe. "Voyages, mesures et instruments: Une nouvelle expérience du monde au siècle des lumières." *Annales: Économies, Sociétés, Civilisations* 52 (1997): 1115–51.

Bourguet, Marie-Noëlle, Christian Licoppe, and H. Otto Sibum. "Introduction." In Bourguet, Licoppe, and Sibum, *Instruments, Travel, and Science*, 1–19.

———, eds. *Instruments, Travel, and Science: Itineraries of Precision from the Seventeenth to the Twentieth Century.* London: Routledge, 2002.

Bradbrook, William. *History of the Parish of Inkberrow.* Worcester: privately printed, [1903?]

Brayne, Martin. *The Greatest Storm.* Stroud, Gloucestershire: Sutton Publishing, 2002.

Brewer, John. *The Pleasures of the Imagination: English Culture in the Eighteenth Century.* New York: Farrar, Straus and Giroux, 1997.

Brewer, John, and Roy Porter, eds. *Consumption and the World of Goods.* London: Routledge, 1993.

Brunhouse, Robert L. "David Ramsay, 1749–1815: Selections from His Writings." *Transactions of the American Philosophical Society*, n.s., 55, pt. 4 (1965): 1–250.

Bryden, D. J. "Sir Samuel Morland's Account of the Balance Barometer, 1678." *Annals of Science* 32 (1975): 359–68.

Burke, Peter. *Popular Culture in Early Modern Europe.* London: Temple Smith, 1978.

———. *The Art of Conversation.* Ithaca, NY: Cornell University Press, 1993.

Burns, William E. "'Our Lot Is Fallen into an Age of Wonders': John Spencer and the Controversy over Prodigies in the Early Restoration." *Albion* 27 (1995): 237–52.

———. *An Age of Wonders: Prodigies, Politics and Providence in England, 1657–1727.* Manchester: Manchester University Press, 2002.

Bushaway, Bob. *By Rite: Custom, Ceremony and Community in England, 1700–1880.* London: Junction Books, 1982.

Caldwell, Deborah. "Was Katrina a Spiritual Message?" *The Advocate* (Baton Rouge), 3 September 2005, 1-F.

Campbell, Mary Baine. *Wonder and Science: Imagining Worlds in Early Modern Europe.* Ithaca, NY: Cornell University Press, 1999.

Cannon, Susan Faye. *Science in Culture: The Early Victorian Period.* New York: Dawson and Science History Publications, 1978.

Capp, Bernard. *Astrology and the Popular Press: English Almanacs, 1500–1800.* London: Faber and Faber, 1979.

Carrithers, David Wallace. "Introduction." In Montesquieu, *The Spirit of Laws*, 3–88.

———. "The Enlightenment Science of Society." In Fox, Porter, and Wokler, *Inventing Human Science*, 232–70.

Carter, Philip. *Men and the Emergence of Polite Society, Britain 1660–1800.* Harlow, Essex: Longman, 2001.

Cassedy, James H. "Meteorology and Medicine in Colonial America: Beginnings of the Experimental Approach." *Journal of the History of Medicine* 24 (1969): 193–204.

Cassidy, David C. "Meteorology in Mannheim: The Palatine Meteorological Society, 1780–1795." *Sudhoffs Archiv* 69 (1985): 8–25.

Castle, Terry. "The Female Thermometer." *Representations*, no. 17 (Winter 1987): 1–27.

Cathcart, Brian. *Rain.* London: Granta Books, 2002.

Chandler, James, Arnold I. Davidson, and Harry Harootunian, eds. *Questions of Evidence: Proof, Practice, and Persuasion across the Disciplines.* Chicago: University of Chicago Press, 1994.

Clark, William, Jan Golinski, and Simon Schaffer, "Introduction." In Clark, Golinski, and Schaffer, *Sciences in Enlightened Europe*, 3–31.

———, eds. *The Sciences in Enlightened Europe.* Chicago: University of Chicago Press, 1999.

Clubbe, John. "The Tempest-Toss'd Summer of 1816: Mary Shelley's *Frankenstein*." *The Byron Journal* 19 (1991): 26–40.

Colley, Linda. *Britons: Forging the Nation, 1707–1837.* New Haven, CT: Yale University Press, 1992.

Cook, Alan. *Edmond Halley: Charting the Heavens and the Seas.* Oxford: Clarendon Press, 1998.

Corbin, Alain. *The Foul and the Fragrant: Odor and the French Social Imagination.* Translated by Miriam L. Kochan. Leamington Spa: Berg Publishers, 1986.

———. *The Lure of the Sea: The Discovery of the Seaside in the Western World, 1750–1840.* Translated by Jocelyn Phelps. London: Penguin Books, 1994.

Costa, Shelley. "The *Ladies' Diary*: Gender, Mathematics, and Civil Society in Early-Eighteenth-Century England." In *Science and Civil Society*, edited by Lynn K. Nyhart and Thomas H. Broman, *Osiris*, 2nd ser., 17 (2002): 49–73. Chicago: University of Chicago Press.

Cranston, Maurice. *John Locke: A Biography.* Oxford: Oxford University Press, 1985.

Crawforth, M. A. "Evidence from Trade Cards for the Scientific Instrument Industry." *Annals of Science* 42 (1985): 453–554.

Cunningham, Andrew, and Roger French, eds. *The Medical Enlightenment of the Eighteenth Century.* Cambridge: Cambridge University Press, 1990.

Curry, Patrick. *Prophecy and Power: Astrology in Early Modern England.* Princeton, NJ: Princeton University Press, 1989.

———. *A Confusion of Prophets: Victorian and Edwardian Astrology.* London: Collins and Brown, 1992.

Daston, Lorraine. "Marvelous Facts and Miraculous Evidence in Early Modern Europe." In Chandler, Davidson, and Harootunian, *Questions of Evidence,* 243–74.

———. "Enlightenment Fears, Fears of Enlightenment." In Baker and Reill, *What's Left of Enlightenment?* 115–28.

———. "Attention and the Values of Nature in the Enlightenment." In Daston and Vidal, *Moral Authority of Nature,* 100–126.

Daston, Lorraine, and Peter Galison. "The Image of Objectivity." *Representations,* no. 40 (1992): 81–128.

Daston, Lorraine, and Katharine Park. *Wonders and the Order of Nature, 1150–1750.* New York: Zone Books, 1998.

Daston, Lorraine, and Fernando Vidal, eds. *The Moral Authority of Nature.* Chicago: University of Chicago Press, 2004.

Davis, Natalie Zemon. "Proverbial Wisdom and Popular Errors." In Davis, *Society and Culture in Early Modern France,* 227–67. Stanford, CA: Stanford University Press, 1975.

Debus, Allen G. "The Paracelsian Aerial Niter." *Isis* 55 (1964): 43–61.

———. "Key to Two Worlds: Robert Fludd's Weather-Glass." *Annali dell'Instituto e Museo di Storia della Scienza di Firenze* 7 (1982): 109–43.

Delbourgo, James. "Common Sense, Useful Knowledge, and Matters of Fact in the Late Enlightenment: The Transatlantic Career of Perkins's Tractors." *William and Mary Quarterly,* 3rd ser., 61 (2004): 643–84.

DePorte, Michael. *Nightmares and Hobbyhorses: Swift, Sterne, and Augustan Ideas of Madness.* San Marino, CA: Huntington Library, 1974.

Dewhurst, Kenneth. *John Locke (1632–1704), Physician and Philosopher: A Medical Biography.* London: Wellcome Historical Medical Library, 1963.

Dixon, Fred E. "Early Irish Weather Records." In Shields, *Irish Meteorological Service,* 59–61.

Dolan, Brian. "Conservative Politicians, Radical Philosophers and the Aerial Remedy for the Diseases of Civilization." *History of the Human Sciences* 15 (2002): 35–54.

Douglas, Aileen. "Popular Science and the Representation of Women: Fontenelle and After." *Eighteenth Century Life* 18, no. 2 (May 1994): 1–14.

Dror, Otniel. "The Scientific Image of Emotion." *Configurations* 7 (1999): 355–401.

Dufour, Louis. *Météorologie, calendriers et croyances populaires: Les origines magico-religieuses, les dictons.* Paris: Librarie d'Amérique et d'Orient, 1978.

Dundes, Alan. "On Whether Weather 'Proverbs' are Proverbs." *Proverbium* 1 (1984): 39–46.

Easlea, Brian. *Science and Sexual Oppression: Patriarchy's Confrontation with Women and Nature.* London: Weidenfeld and Nicolson, 1981.

Eisenstadt, Peter. "The Weather and Weather Forecasting in Colonial America." Ph.D. diss., New York University, 1990.

Evelyn, John. *Diary and Correspondence of John Evelyn, F.R.S.* Edited by William Bray. London: George Routledge and Sons, n.d.

Eze, Emmanuel C. "Hume, Race, and Human Nature." *Journal of the History of Ideas* 61 (2000): 691–98.

Fara, Patricia. "Lord Derwentwater's Lights: Prediction and the Aurora Polaris." *Journal of the History of Astronomy* 27 (1996): 239–58.

———. *Pandora's Breeches: Women, Science, and Power in the Enlightenment.* London: Pimlico, 2004.

Feldman, Theodore S. "Late Enlightenment Meteorology." In Frängsmyr, Heilbron, and Rider, *Quantifying Spirit*, 143–77.

———. "The Ancient Climate in the Eighteenth and Early Nineteenth Century." In *Science and Nature: Essays in the History of the Environmental Sciences*, edited by Michael Shortland, 23–40. Chalfont St. Giles, Buckinghamshire: British Society for the History of Science, 1993.

Fleming, James Rodger. *Meteorology in America, 1800–1870.* Baltimore: Johns Hopkins University Press, 1990.

———. *Historical Perspectives on Climate Change.* New York: Oxford University Press, 1998.

Fortune, Brandon Brame. *Franklin and His Friends: Portraying the Man of Science in Eighteenth-Century America.* With Deborah J. Warner. Washington, DC: Smithsonian / National Portrait Gallery, 1999.

Foucault, Michel. "What Is Enlightenment?" In Rabinow, *The Foucault Reader*, 32–50.

———. "On the Genealogy of Ethics: An Overview of Work in Progress." In Rabinow, *The Foucault Reader*, 340–72.

Fox, Adam. *Oral and Literate Culture in England, 1500–1700.* Oxford: Clarendon Press, 2000.

Fox, Christopher, Roy Porter, and Robert Wokler, eds., *Inventing Human Science: Eighteenth-Century Domains.* Berkeley and Los Angeles: University of California Press, 1995.

Frängsmyr, Carl. *Klimat och karaktär: Naturen och människan i sent svenskt 1700-tal.* Uppsala: Natur och Kultur, 2000.

Frängsmyr, Tore, J. L. Heilbron, and Robin Rider, eds. *The Quantifying Spirit in the Eighteenth Century.* Berkeley and Los Angeles: University of California Press, 1990.

Frank, Robert G., Jr. *Harvey and the Oxford Physiologists: Scientific Ideas and Social Interaction.* Berkeley and Los Angeles: University of California Press, 1978.

———. "Medicine." In *The History of the University of Oxford*, vol. 4, *Seventeenth-Century Oxford*, edited by Nicholas Tyacke, 359–448. Oxford: Clarendon Press, 1997.

Freshfield, Douglas W., and Henry F. Montagnier. *The Life of Horace Benedict de Saussure.* London: Edward Arnold, 1920.

Freud, Sigmund. *Civilization, Society and Religion: Group Psychology, Civilization and Its Discontents and Other Works.* Penguin Freud Library, vol. 12. London: Penguin Books, 1991.

Friedman, Robert Marc. *Appropriating the Weather: Vilhelm Bjerknes and the Construction of a Modern Meteorology.* Ithaca, NY: Cornell University Press, 1989.

Frisinger, H. Howard. *The History of Meteorology to 1800*. New York: Science History Publications, 1977.

Furbank, P. N., and W. R. Owens. *A Critical Bibliography of Daniel Defoe*. London: Pickering and Chatto, 1998.

Galison, Peter. *Image and Logic: A Material Culture of Microphysics*. Chicago: University of Chicago Press, 1997.

Galtier, Charles. *Météorologie populaire dans la France ancienne: La Provence, empire du soleil et royaume des vents*. Le Coteau: Editions Horvath, 1984.

Garriott, Edward B. *Weather Folk-Lore and Local Weather Signs*. Washington, DC: Government Printing Office, 1903.

Geneva, Ann. *Astrology and the Seventeenth-Century Mind: William Lilly and the Language of the Stars*. Manchester: Manchester University Press, 1995.

Gerbi, Antonello. *The Dispute of the New World: The History of a Polemic*. Translated by Jeremy Moyle. Pittsburgh: University of Pittsburgh Press, 1973.

Gidal, Eric. "Civic Melancholy: English Gloom and French Enlightenment." *Eighteenth-Century Studies* 37 (2003): 23–45.

Giroux, Jacqueline. "Genèse de la météorologie scientifique dans les milieux de l'Académie de Dijon au XVIIIe siècle." *Mémoires de l'Académie de Dijon* 125 (1981–82): 135–55.

Glacken, Clarence J. *Traces on the Rhodian Shore: Nature and Culture in Western Thought from Ancient Times to the End of the Eighteenth Century*. Berkeley and Los Angeles: University of California Press, 1967.

Godwin, Jocelyn. *Robert Fludd: Hermetic Philosopher and Surveyor of Two Worlds*. London: Thames and Hudson, 1979.

Golinski, Jan. *Science as Public Culture: Chemistry and Enlightenment in Britain, 1760–1820*. Cambridge: Cambridge University Press, 1992.

———. "Barometers of Change: Meteorological Instruments as Machines of Enlightenment." In Clark, Golinski, and Schaffer, *Sciences in Enlightened Europe*, 69–93.

———. "'Exquisite Atmography': Theories of the World and Experiences of the Weather in a Diary of 1703." *British Journal for the History of Science* 34 (2001): 149–71.

———. "The Care of the Self and the Masculine Birth of Science." *History of Science* 40 (2002): 125–45.

———. "Chemistry." In *The Cambridge History of Science*, vol. 4, *Eighteenth-Century Science*, edited by Roy Porter, 375–96. Cambridge: Cambridge University Press, 2003.

———. "Time, Talk, and the Weather in Eighteenth-Century Britain." In Strauss and Orlove, *Weather, Climate, Culture*, 17–38.

———. *Making Natural Knowledge: Constructivism and the History of Science*. New ed. Chicago: University of Chicago Press, 2005.

Goodison, Nicholas. *English Barometers, 1680–1860: A History of Domestic Barometers and Their Makers and Retailers*. Woodbridge, Suffolk: Antique Collectors' Club, 1977.

Gould, Stephen Jay. "Worm for a Century, and All Seasons." In Gould, *Hen's Teeth and Horse's Toes: Further Reflections in Natural History*, 120–33. Harmondsworth, Middlesex: Penguin Books, 1984.

Grattan, John, and Mark Brayshay. "An Amazing and Portentous Summer: Environmental and Social Responses in Britain to the 1783 Eruption of an Iceland Volcano." *The Geographical Journal* 161 (1995): 125–34.

Greig, J. Y. T., ed. *The Letters of David Hume.* 2 vols. Oxford: Clarendon Press, 1932.

Grove, Richard H. *Green Imperialism: Colonial Expansion, Tropical Island Edens, and the Origins of Environmentalism, 1600–1860.* Cambridge: Cambridge University Press, 1995.

Guerlac, Henry. "The Poets' Nitre." *Isis* 45 (1954): 243–55.

Guerrini, Anita. "Chemistry Teaching at Oxford and Cambridge, circa 1700." In *Alchemy and Chemistry in the 16th and 17th Centuries*, edited by P. Rattansi and A. Clericuzio, 183–99. Dordrecht: Kluwer, 1994.

———. *Obesity and Depression in the Enlightenment: The Life and Times of George Cheyne.* Norman: University of Oklahoma Press, 2000.

Gunther, Robert T. *Early Science in Oxford.* 14 vols. Oxford: for the Subscribers, 1923–45.

Habermas, Jürgen. *The Structural Transformation of the Public Sphere: An Inquiry into a Category of Bourgeois Society.* Translated by Thomas Burger with the assistance of Frederick Lawrence. Cambridge, MA: MIT Press, 1991.

Hall, David D. *Worlds of Wonder, Days of Judgment: Popular Religious Belief in Early New England.* Cambridge, MA: Harvard University Press, 1990.

Hamblyn, Richard. *The Invention of Clouds: How an Amateur Meteorologist Forged the Language of the Skies.* New York: Farrar, Straus and Giroux, 2001.

———. "Introduction." In Defoe, *The Storm* (2005), x–xl.

Hamlin, Christopher. "Providence and Putrefaction: Victorian Sanitarians and the Natural Theology of Health and Disease." In *Energy and Entropy: Science and Culture in Victorian Britain*, edited by Patrick Brantlinger, 93–123. Bloomington: Indiana University Press, 1989.

———. "Environmental Sensibility in Edinburgh, 1839–1840: The 'Fetid Irrigation' Controversy." *Journal of Urban History* 20 (1994): 311–39.

———. *Public Health and Social Justice in the Age of Chadwick: Britain, 1800–1854.* Cambridge: Cambridge University Press, 1998.

Hannaway, Caroline. "Environment and Miasmata." In *Companion Encyclopedia of the History of Medicine*, edited by W. F. Bynum and Roy Porter, 1:292–308. London: Routledge, 1993.

Harley, Trevor A. "Nice Weather for the Time of Year: The British Obsession with the Weather." In Strauss and Orlove, *Weather, Climate, Culture*, 103–18.

Harrison, Mark. "'The Tender Frame of Man': Disease, Climate, and Racial Difference in India and the West Indies." *Bulletin of the History of Medicine* 70 (1996): 68–93.

———. *Climates and Constitutions: Health, Race, Environment, and British Imperialism in India, 1600–1850.* Oxford: Oxford University Press, 1999.

———. "From Medical Astrology to Medical Astronomy: Sol-Lunar and Planetary Theories of Disease in British Medicine, c.1700–1850." *British Journal for the History of Science* 33 (2000): 25–48.

Harvey, William F. *John Rutty of Dublin, Quaker Physician*. The Lister Lecture, delivered before the Quaker Medical Society, 3 November 1933. Reprinted from *The Friends' Quarterly Examiner*, January–April 1934.

Hay, Andrew. *The Diary of Andrew Hay of Craignethan, 1659–1660*. Edited by Alexander George Reid. Edinburgh: Scottish History Society, 1901.

Head, Mrs. Henry. *The Weather Calendar, or a Record of the Weather for Every Day of the Year*. Oxford: Oxford University Press, 1917.

Hearne, Thomas. *Remarks and Collections*. 11 vols., Oxford: Clarendon Press for the Oxford Historical Society, 1885–1918.

Heffernan, Michael. "Historical Geographies of the Future: Three Perspectives from France, 1750–1825." In Livingstone and Withers, *Geography and Enlightenment*, 125–64.

Heninger, S. K. *A Handbook of Renaissance Meteorology*. Durham, NC: Duke University Press, 1960.

Heyd, Michael. "Medical Discourse in Religious Controversy: The Case of the Critique of 'Enthusiasm' on the Eve of the Enlightenment." *Science in Context* 8 (1995): 133–57.

Hobbs, William. *The Earth Generated and Anatomized*. Edited by Roy Porter. London: British Museum (Natural History), 1981.

Hunt, Robert, and Ruth Jackson. *More about Inkberrow*. Inkberrow, Worcestershire: privately printed, [1976?]

———. *Inkberrow Ways*. Inkberrow, Worcestershire: privately printed, n.d.

Hunter, Lynette, and Sarah Hutton, eds. *Women, Science, and Medicine, 1500–1700: Mothers and Sisters of the Royal Society*. Stroud, Gloucestershire: Sutton Publishing, 1997.

Hunter, Michael, and Annabel Gregory, eds. *An Astrological Diary of the Seventeenth Century: Samuel Jeake of Rye, 1652–1699*. Oxford: Clarendon Press, 1988.

Israel, Jonathan I., and Geoffrey Parker. "Of Providence and Protestant Winds: The Spanish Armada of 1588 and the Dutch Armada of 1688." In *The Anglo-Dutch Moment: Essays on the Glorious Revolution and Its World Impact*, edited by Jonathan I. Israel, 335–63. Cambridge: Cambridge University Press, 1991.

Janković, Vladimir. "Ideological Crests versus Empirical Troughs: John Herschel's and William Radcliffe Birt's Research on Atmospheric Waves, 1843–50." *British Journal for the History of Science* 31 (1998): 21–40.

———. *Reading the Skies: A Cultural History of English Weather, 1650–1820*. Chicago: University of Chicago Press, 2000.

———. "Meteorology." In *Encyclopedia of the Enlightenment*, edited by Alan Charles Kors, 3:70–71. Oxford: Oxford University Press, 2003.

Jardine, Lisa. *The Curious Life of Robert Hooke, the Man Who Measured London*. New York: HarperCollins, 2003.

Jardine, Nick. "Whigs and Stories: Herbert Butterfield and the Historiography of Science." *History of Science* 41 (2003): 125–40.

Jenks, Stuart. "Astrometeorology in the Middle Ages." *Isis* 74 (1983): 185–210.

Jenner, Mark. "Bathing and Baptism: Sir John Floyer and the Politics of Cold Bathing." In *Refiguring Revolutions: Aesthetics and Politics from the English Revolution to the Romantic Revolution*, edited by Kevin Sharpe and Steven N. Zwicker, 197–216. Berkeley and Los Angeles: University of California Press, 1998.

Johnson, Samuel. *The Yale Edition of the Works of Samuel Johnson*. Edited by W. J. Bate, John M. Bullitt, and M. F. Powell. 18 vols. New Haven, CT: Yale University Press, 1958–.

Jones, Michael Owen. "Climate and Disease: The Traveler Describes America." *Bulletin of the History of Medicine* 41 (1967): 254–66.

Jordanova, Ludmilla. "Earth Science and Environmental Medicine: The Synthesis of the Late Enlightenment." In *Images of the Earth*, edited by Ludmilla Jordanova and Roy Porter, 2nd ed., 127–51. Chalfont St. Giles, Buckinghamshire: British Society for the History of Science, 1997.

Kington, John, ed. *The Weather Journals of a Rutland Squire: Thomas Barker of Lyndon Hall*. Oakham, Rutland: Rutland Record Society, 1988.

———. *The Weather of the 1780s over Europe*. Cambridge: Cambridge University Press, 1988.

Kiple, Kenneth F., and Kriemhild Coneè Ornelas. "Race, War and Tropical Medicine in the Eighteenth-Century Caribbean." In Arnold, *Warm Climates and Western Medicine*, 65–79.

Klein, Lawrence E. *Shaftesbury and the Culture of Politeness: Moral Discourse and Cultural Politics in Early Eighteenth-Century England*. Cambridge: Cambridge University Press, 1994.

———. "Coffeehouse Civility, 1660–1714: An Aspect of Post-Courtly Culture in England." *Huntington Library Quarterly* 59 (1996): 30–51.

———. "Enlightenment as Conversation." In Baker and Reill, *What's Left of Enlightenment?* 148–66.

Kuhn, Thomas S. "Mathematical versus Experimental Traditions in the Development of Physical Science." In Kuhn, *The Essential Tension: Selected Studies in Scientific Tradition and Change*, 31–65. Chicago: University of Chicago Press, 1977.

Kupperman, Karen Ordahl. "The Puzzle of the American Climate in the Early Colonial Period." *American Historical Review* 87 (1982): 1262–89.

———. "Fear of Hot Climates in the Anglo-American Colonial Experience." *William and Mary Quarterly*, 3rd ser., 41 (1984): 213–40.

Langford, Paul. *A Polite and Commercial People: England, 1727–1783*. Oxford: Clarendon Press, 1989.

Latour, Bruno. *We Have Never Been Modern*. Translated by Catherine Porter. Cambridge, MA: Harvard University Press, 1993.

Lawrence, Christopher, and Steven Shapin, eds. *Science Incarnate: Historical Embodiments of Natural Knowledge*. Chicago: University of Chicago Press, 1998.

Leigh, R. A. A. *The Eton College Register, 1698–1752*. Eton: Spottiswood, Ballantyne, and Co., 1927.

Levere, T. H., and G. L'E. Turner. *Discussing Chemistry and Steam: The Minutes of a Coffee House Philosophical Society, 1780–1787*. Oxford: Oxford University Press, 2002.

Licoppe, Christian. "The Project for a Map of Languedoc in Eighteenth-Century France at the Contested Intersection between Astronomy and Geography: The Problem of Co-ordination between Philosophers, Instruments and Observations as a Keystone of Modernity." In Bourguet, Licoppe, and Sibum, *Instruments, Travel, and Science*, 51–74.

Livingstone, David N. *Putting Science in Its Place: Geographies of Scientific Knowledge*. Chicago: University of Chicago Press, 2003.

Livingstone, David N., and Charles W. J. Withers, eds. *Geography and Enlightenment*. Chicago: University of Chicago Press, 1999.

Lloyd, G. E. R., ed. *Hippocratic Writings*. Harmondsworth, Middlesex: Penguin Books, 1978.

Mabey, Richard. *Gilbert White: A Biography of the Naturalist and Author of "The Natural History of Selborne."* London: Pimlico, 1999.

MacDonald, Michael. *Mystical Bedlam: Madness, Anxiety, and Healing in Seventeenth-Century England*. Cambridge: Cambridge University Press, 1981.

Macfarlane, Alan. *The Family Life of Ralph Josselin, a Seventeenth-Century Clergyman: An Essay in Historical Anthropology*. Cambridge: Cambridge University Press, 1970.

———, ed. *The Diary of Ralph Josselin, 1616–1683*. Oxford: Oxford University Press, 1976.

Malcolmson, Robert W. *Popular Recreations in English Society, 1700–1850*. Cambridge: Cambridge University Press, 1988.

Manley, Gordon. "The Weather and Diseases: Some Eighteenth-Century Contributions to Observational Meteorology." *Notes and Records of the Royal Society of London* 9 (1952): 300–307.

Manuel, Frank E. *Isaac Newton, Historian*. Cambridge: Cambridge University Press, 1963.

Martin, Luther H., Huck Gutman, and Patrick H. Hutton, eds. *Technologies of the Self: A Seminar with Michel Foucault*. Amherst: University of Massachusetts Press, 1988.

Matthews, William. "Polite Speech in the Eighteenth Century." *English: The Magazine of the English Association* 1 (1936–37): 493–511.

———. *British Diaries: An Annotated Bibliography of British Diaries Written between 1442 and 1942*. Berkeley and Los Angeles: University of California Press, 1950.

McEvoy, John G. "Joseph Priestley, 'Aerial Philosopher': Metaphysics and Methodology in Priestley's Thought, 1772–1781." *Ambix* 25 (1978): 1–55, 93–116, 153–75; 26 (1979): 16–38.

McKeon, Michael. *The Origins of the English Novel, 1600–1740*. Baltimore: Johns Hopkins University Press, 1987.

McWilliams, Brendan. "The Kingdom of the Air: The Progress of Meteorology." In *Nature in Ireland: A Scientific and Cultural History*, edited by John Wilson Foster, 115–32. Dublin: Lilliput Press, 1997.

Megill, Allan, ed. *Rethinking Objectivity*. Durham, NC: Duke University Press, 1994.

Merchant, Carolyn. *The Death of Nature: Women, Ecology, and the Scientific Revolution*. San Francisco: Harper and Row, 1980.

Meyer, William B. *Americans and Their Weather*. Oxford: Oxford University Press, 2000.

Middleton, W. E. Knowles. *The History of the Barometer*. Baltimore: Johns Hopkins University Press, 1964.

————. *A History of the Thermometer and Its Uses in Meteorology*. Baltimore: Johns Hopkins University Press, 1966.

Midgley, Mary. *Science and Poetry*. London: Routledge, 2001.

Miller, Charles A. *Jefferson and Nature: An Interpretation*. Baltimore: Johns Hopkins University Press, 1988.

Monmonier, Mark. *Air Apparent: How Meteorologists Learned to Map, Predict, and Dramatize Weather*. Chicago: University of Chicago Press, 1999.

Morrell, Jack, and Arnold Thackray. *Gentlemen of Science: Early Years of the British Association for the Advancement of Science*. Oxford: Clarendon Press, 1981.

Mullan, John. *Sentiment and Sociability: The Language of Feeling in the Eighteenth Century*. Oxford: Oxford University Press, 1988.

————. "Gendered Knowledge, Gendered Minds: Women and Newtonianism, 1690–1760." In *A Question of Identity: Women, Science, and Literature*, edited by Marina Benjamin, 41–56. New Brunswick, NJ: Rutgers University Press, 1993.

Mulligan, Lotte. "Self-Scrutiny and the Study of Nature: Robert Hooke's Diary as Natural History." *Journal of British Studies* 35 (1996): 311–42.

Myer, Michael Grosvenor. "Talking of the Weather: A Note on Manners in Jane Austen." *Notes and Queries* 241 (new ser. 43) (1996): 418.

Nielsen, Rosemary M., and Robert H. Solomon. "Writing the Bodies of Water: The Clash of the Lasting and the Catastrophic in the *Odes* of Horace." In *Climate and Literature: Reflections of Environment*, edited by Janet Pérez and Wendell Aycock, 7–18. Lubbock: Texas Tech University Press, 1995.

North, Roger. *Notes of Me: The Autobiography of Roger North*. Edited by Peter Millard. Toronto: University of Toronto Press, 2000.

Nussbaum, Felicity A. *The Autobiographical Subject: Gender and Ideology in Eighteenth-Century England*. Baltimore: Johns Hopkins University Press, 1989.

Obelkevich, James. "Proverbs and Social History." In *The Social History of Language*, edited by Peter Burke and Roy Porter, 43–72. Cambridge: Cambridge University Press, 1987.

Oldenburg, Henry. *The Correspondence of Henry Oldenburg*. Edited by A. R. Hall and Marie Boas Hall. 13 vols. Madison: University of Wisconsin Press; London: Mansell; London: Taylor and Francis, 1965–86.

Oliver, J. "William Borlase's Contribution to Eighteenth-Century Meteorology and Climatology." *Annals of Science* 25 (1969): 275–317.

Orlove, Ben. "How People Name Seasons." In Strauss and Orlove, *Weather, Climate, Culture*, 121–40.

Outram, Dorinda. "The Enlightenment Our Contemporary." In Clark, Golinski, and Schaffer, *Sciences in Enlightened Europe*, 32–40.

————. "On Being Perseus: New Knowledge, Dislocation, and Enlightenment Exploration." In Livingstone and Withers, *Geography and Enlightenment*, 281–94.

Overnell, R. F. *The Ashmolean Museum, 1683–1894*. Oxford: Clarendon Press, 1986.

The Oxford Classical Dictionary. Edited by Simon Hornblower and Antony Spawforth. Oxford: Oxford University Press, 1996.

Passy, Charles. "Tragedies of Nature, Terror Leave Sense of Dread." *Palm Beach Post* (Florida), 12 September 2005, 1A.

Payne, Harry C. "Elite versus Popular Mentality in the Eighteenth Century." *Studies in Eighteenth-Century Culture* 8 (1979): 3–32.

Perkins, Maureen. *Visions of the Future: Almanacs, Time, and Cultural Change, 1775–1870*. Oxford: Clarendon Press, 1996.

Pickering, Michael. "The Four Angels of the Earth: Popular Cosmology in a Victorian Village." *Southern Folklore Quarterly* 45 (1981): 1–18.

Pickstone, John. "Dearth, Dirt and Fever Epidemics: Rewriting the History of British 'Public Health,' 1780–1850." In *Epidemics and Ideas: Essays on the Historical Perception of Pestilence*, edited by Terence Ranger and Paul Slack, 125–48. Cambridge: Cambridge University Press, 1992.

Poole, Robert. " 'Give Us Our Eleven Days!' Calendar Reform in Eighteenth-Century England." *Past and Present*, no. 149 (1995): 95–139.

———. *Time's Alteration: Calendar Reform in Early Modern England*. London: UCL Press, 1998.

Popkin, Richard H. "The Philosophical Basis of Eighteenth-Century Racism." In *Racism in the Eighteenth Century*, edited by Harold E. Pagliaro, *Studies in Eighteenth-Century Culture* 3 (1973): 245–62. Cleveland: Case Western Reserve University Press.

Porter, Roy. "The Rage of Party: A Glorious Revolution in English Psychiatry?" *Medical History* 27 (1983): 35–50.

———. "Lay Medical Knowledge in the Eighteenth Century: The Evidence of the *Gentleman's Magazine*." *Medical History* 29 (1985): 138–68.

———. "Introduction." In Cheyne, *The English Malady* (1991), ix–li.

———. *Doctor of Society: Thomas Beddoes and the Sick Trade in Late-Enlightenment England*. London: Routledge, 1992.

———. "Consumption: Disease of the Consumer Society?" In Brewer and Porter, *Consumption and the World of Goods*, 58–81.

———. *The Creation of the Modern World: The Untold Story of the British Enlightenment*. New York: W. W. Norton, 2000.

———. *Flesh in the Age of Reason*. New York: W. W. Norton, 2003.

Porter, Roy, and Mikuláš Teich, eds. *The Enlightenment in National Context*. Cambridge: Cambridge University Press, 1981.

Rabinow, Paul, ed. *The Foucault Reader*. New York: Pantheon Books, 1984.

Raines, Robert. *Marcellus Laroon*. London: Routledge and Kegan Paul, 1966.

Reed, Arden. *Romantic Weather: The Climates of Coleridge and Baudelaire*. Hanover, NH: University Press of New England, 1983.

Reedy, Gerard, S.J. "Mystical Politics: The Imagery of Charles II's Coronation." In *Studies in Change and Revolution: Aspects of English Intellectual History, 1640–1800*, edited by Paul J. Korshin, 19–42. Menston, Yorkshire: Scolar Press, 1973.

Rees, Ronald. "Under the Weather: Climate and Disease, 1700–1900." *History Today* 46, no. 1 (January 1996): 35–41.

Riley, James C. *The Eighteenth-Century Campaign to Avoid Disease*. Basingstoke: Macmillan, 1987.

Risk Management Solutions. *December 1703 Windstorm: 300-Year Retrospective*. http://www.rms.com/Publications/1703_Windstorm.pdf (accessed 18 February 2004).

Riskin, Jessica. *Science in the Age of Sensibility: The Sentimental Empiricists of the French Enlightenment.* Chicago: University of Chicago Press, 2002.

Romani, Roberto. *National Character and Public Spirit in Britain and France, 1750–1914.* Cambridge: Cambridge University Press, 2002.

Ross, Andrew. *Strange Weather: Culture, Science, and Technology in the Age of Limits.* London: Verso, 1991.

Rossi, Paolo. *The Dark Abyss of Time: The History of the Earth and the History of Nations from Hooke to Vico.* Translated by Lydia G. Cochrane. Chicago: University of Chicago Press, 1984.

Rothstein, Edward. "Seeking Justice, of Gods or the Politicians." *New York Times,* 8 September 2005, E-1.

Rusnock, Andrea A. "Hippocrates, Bacon, and Medical Meteorology at the Royal Society, 1700–1750." In *Reinventing Hippocrates,* edited by David Cantor, 136–53. Aldershot, Hampshire: Ashgate Press, 2002.

———. *Vital Accounts: Quantifying Health and Population in Eighteenth-Century England and France.* Cambridge: Cambridge University Press, 2002.

———, ed. *The Correspondence of James Jurin (1684–1750), Physician and Secretary to the Royal Society.* Amsterdam: Rodopi, 1996.

Sarasohn, Lisa Tunick. "Epicureanism and the Creation of a Privatist Ethic in Early Seventeenth-Century France." In *Atoms, Pneuma, and Tranquility: Epicurean and Stoic Themes in European Thought,* edited by Margaret J. Osler, 175–95. Cambridge: Cambridge University Press, 1991.

Sargent, Frederick, II. *Hippocratic Heritage: A History of Ideas about Weather and Human Health.* New York: Pergamon Press, 1982.

Schaffer, Simon. "Measuring Virtue: Eudiometry, Enlightenment and Pneumatic Medicine." In Cunningham and French, *Medical Enlightenment,* 281–318.

Schechner, Sara J. *Comets, Popular Culture, and the Birth of Modern Cosmology.* Princeton, NJ: Princeton University Press, 1997.

Schmidt, James, ed. *What Is Enlightenment? Eighteenth-Century Answers and Twentieth-Century Questions.* Berkeley and Los Angeles: University of California Press, 1996.

Schmidt, James, and Thomas E. Wartenberg. "Foucault's Enlightenment: Critique, Revolution, and the Fashioning of the Self." In *Critique and Power: Recasting the Foucault/Habermas Debate,* edited by Michael Kelly, 283–314. Cambridge, MA: MIT Press, 1994.

Schofield, Robert E. *Mechanism and Materialism: British Natural Philosophy in an Age of Reason.* Princeton, NJ: Princeton University Press, 1970.

———. *The Enlightenment of Joseph Priestley: A Study of His Life and Work from 1733 to 1773.* University Park: Pennsylvania State University Press, 1997.

———. *The Enlightened Joseph Priestley: A Study of His Life and Work from 1773 to 1804.* University Park: Pennsylvania State University Press, 2004.

Schove, D. J. "Weather." In *The Diary of Samuel Pepys,* vol. 10, *Companion,* edited by R. Latham, 470–71. London: Bell and Hyman, 1983.

———, and David Reynolds. "Weather in Scotland, 1659–1660: The Diary of Andrew Hay." *Annals of Science* 30 (1973): 165–78.

Schwartz, Hillel. *Knaves, Fools, Madmen, and That Subtle Effluvium: A Study of the Opposition to the French Prophets in England, 1706–1710.* Gainesville: University Presses of Florida, 1978.

Scott, D. F. S., ed. *Luke Howard (1772–1864): His Correspondence with Goethe and His Continental Journey of 1816.* York: William Sessions, 1976.

Seabrook, John. "Selling the Weather." *The New Yorker*, 3 April 2000, 44.

Serres, Michel. *The Natural Contract.* Translated by Elizabeth MacArthur and William Paulson. Ann Arbor: University of Michigan Press, 1995.

———. *Genesis.* Translated by Geneviève James and James Nielson. Ann Arbor: University of Michigan Press, 1995.

———. *The Birth of Physics.* Translated by Jack Hawkes. Edited by David Webb. Manchester: Clinamen Press, 2000.

Shapin, Steven. "Who Was Robert Hooke?" In *Robert Hooke: New Studies*, edited by Michael Hunter and Simon Schaffer, 253–85. Woodbridge, Suffolk: Boydell Press, 1989.

———. *A Social History of Truth: Civility and Science in Seventeenth-Century England.* Chicago: University of Chicago Press, 1994.

———. "The Philosopher and the Chicken: On the Dietetics of Disembodied Knowledge." In Lawrence and Shapin, *Science Incarnate*, 21–50.

———. "Proverbial Economies: How an Understanding of Some Linguistic and Social Features of Common Sense Can Throw Light on More Prestigious Bodies of Knowledge, Science for Example." *Social Studies of Science* 31 (2001): 731–69.

Shapiro, Barbara J. *A Culture of Fact: England, 1550–1720.* Ithaca, NY: Cornell University Press, 2000.

Shaw, Napier. *Meteorology in History.* Vol. 1 of *Manual of Meteorology.* With Elaine Austin. Cambridge: Cambridge University Press, 1926.

Sherbo, Arthur. "The English Weather, *The Gentleman's Magazine*, and the Brothers White." *Archives of Natural History* 12 (1985): 23–29.

Sherman, Stuart. *Telling Time: Clocks, Diaries, and English Diurnal Form, 1660–1785.* Chicago: University of Chicago Press, 1996.

Shields, Lisa. "Popular Weather Lore in Ireland." In Shields, *Irish Meteorological Service*, 56–58.

———, ed. *The Irish Meteorological Service: The First Fifty Years, 1936–1986.* Dublin: Stationery Office, 1987.

Shteir, Ann. "'Conversable Rather Than Scientific': Women and Late Enlightenment Science Culture in England." *Studies on Voltaire and the Eighteenth Century* 304 (1992): 768–72.

Simcock, A. V. *The Ashmolean Museum and Oxford Science, 1683–1983.* Oxford: Museum of the History of Science, 1984.

Simon, Gérard. *Kepler, astronome astrologue.* Paris: Gallimard, 1979.

Singer, Dorothea Wade. "Sir John Pringle and His Circle." *Annals of Science* 6 (1948–50): 127–80, 229–61.

Slatkin, Laura M. "Measuring Authority, Authoritative Measures: Hesiod's *Works and Days*." In Daston and Vidal, *Moral Authority of Nature*, 25–49.

Sloane, Eric. *Folklore of American Weather.* New York: Duell, Sloan, and Pearce, 1963.

Sluga, Hans. "Heidegger and the Critique of Reason." In Baker and Reill, *What's Left of Enlightenment?* 50–70.

Smart, Christopher. "Jubilate Agno." In Smart, *Selected Poems*, edited by Karina Williamson and Marcus Walsh. London: Penguin Books, 1990.

Smith, Roger. "The Language of Human Nature." In Fox, Porter, and Wokler, *Inventing Human Science*, 88–111.

———. "Self-Reflection and the Self." In *Rewriting the Self: Histories from the Middle Ages to the Present*, edited by Roy Porter, 49–57. London: Routledge, 1997.

Smitten, Jeffrey. "Impartiality in Robertson's *History of America*." *Eighteenth-Century Studies* 19 (1985): 56–77.

Sorrenson, Richard. "The Ship as a Scientific Instrument in the Eighteenth Century." In *Science in the Field*, edited by Henricka Kuklick and Robert E. Kohler, *Osiris*, 2nd ser., 11 (1996): 221–36. Chicago: University of Chicago Press.

Spadafora, David. *The Idea of Progress in Eighteenth-Century Britain*. New Haven, CT: Yale University Press, 1990.

Spitzer, Leo. "Milieu and Ambiance." In Spitzer, *Essays in Historical Semantics*, 179–316. New York: S. F. Vanni, 1948.

Spurr, John. "'Virtue, Religion and Government': The Anglican Uses of Providence." In *The Politics of Religion in Restoration England*, edited by Tim Harris, Paul Seaward, and Mark Goldie, 29–47. Oxford: Basil Blackwell, 1990.

Stearns, Raymond Phineas. *Science in the British Colonies of America*. Urbana: University of Illinois Press, 1970.

Stewart, Larry. *The Rise of Public Science: Rhetoric, Technology, and Natural Philosophy in Newtonian Britain, 1660–1750*. Cambridge: Cambridge University Press, 1992.

Stommel, Henry, and Elizabeth Stommel. "The Year without a Summer." *Scientific American* 240 (June 1979): 176–86.

Strauss, Sarah, and Benjamin S. Orlove. "Up in the Air: The Anthropology of Weather and Climate." In Strauss and Orlove, *Weather, Climate, Culture*, 3–14.

———, eds. *Weather, Climate, Culture*. Oxford: Berg Publishers, 2003.

Sutton, Geoffrey V. *Science for a Polite Society: Gender, Culture, and the Demonstration of Enlightenment*. Boulder, CO: Westview Press, 1995.

Svenvold, Mark. "Look Dear—More Catastrophes!" *Washington Post*, 6 November 2005, B-1.

Taub, Liba. *Ancient Meteorology*. London: Routledge, 2003.

Taylor, E. G. R. *The Mathematical Practitioners of Tudor and Stuart England*. Cambridge: for the Institute of Navigation at the University Press, 1954.

Taylor, F. Sherwood. "The Origin of the Thermometer." *Annals of Science* 5 (1942): 129–56.

Thomas, Keith. *Religion and the Decline of Magic*. Harmondsworth, Middlesex: Penguin Books, 1973.

———. *Man and the Natural World: Changing Attitudes in England, 1500–1800*. Harmondsworth, Middlesex: Penguin Books, 1983.

Thompson, E. P. "Time, Work-Discipline and Industrial Capitalism." *Past and Present*, no. 38 (1967): 56–97.

Todd, Dennis. *Imagining Monsters: Miscreations of the Self in Eighteenth-Century England*. Chicago: University of Chicago Press, 1995.

Turney, Jon. *Frankenstein's Footsteps: Science, Genetics and Popular Culture*. New Haven, CT: Yale University Press, 1998.

Udías, Agustín. "Meteorology in the Observatories of the Society of Jesus." *Archivium Historicum Societatis Iesu* 65 (1996): 157–70.

Valenčius, Conevery Bolton. *The Health of the Country: How American Settlers Understood Themselves and Their Land*. New York: Basic Books, 2002.

Van Sant, Ann Jessie. *Eighteenth-Century Sensibility and the Novel: The Senses in Social Context*. Cambridge: Cambridge University Press, 1993.

Vermeir, Koen. "The 'Physical Prophet' and the Powers of the Imagination; Part I: A Case-Study on Prophecy, Vapours and the Imagination (1685–1710)." *Studies in History and Philosophy of Biological and Biomedical Sciences* 35 (2004): 561–91.

———. "The 'Physical Prophet' and the Powers of the Imagination; Part II: A Case-Study on Dowsing and the Naturalisation of the Moral, 1685–1710." *Studies in History and Philosophy of Biological and Biomedical Sciences* 36 (2005): 1–24.

The Victoria History of the County of Oxford. 13 vols. London: for the University of London Institute for Historical Research by Oxford University Press, 1939–96.

Vila, Anne C. *Enlightenment and Pathology: Sensibility in the Literature and Medicine of Eighteenth-Century France*. Baltimore: Johns Hopkins University Press, 1998.

Wagner, Peter. "'An Entirely New Object of Consciousness, of Volition, of Thought': The Coming into Being and (Almost) Passing Away of 'Society' as a Scientific Object." In *Biographies of Scientific Objects*, edited by Lorraine Daston, 132–57. Chicago: University of Chicago Press, 2000.

Walters, Alice N. "Tools of Enlightenment: The Material Culture of Science in Eighteenth-Century England." Ph.D. diss., University of California, Berkeley, 1992.

———. "Conversation Pieces: Science and Politeness in Eighteenth-Century England." *History of Science* 35 (1997): 121–54.

Ward, Donald J. "Weather Signs and Weather Magic: Some Ideas on Causality in Popular Belief." *Pacific Coast Philology* 3 (1968): 67–72.

Webb, Nicholas. "Representations of the Seasons in Early Nineteenth-Century England." Ph.D. diss., University of York 1998.

Webster, Tom. "Writing to Redundancy: Approaches to Spiritual Journals and Early-Modern Spirituality." *The Historical Journal* 39 (1996): 33–56.

Wheeler, Dennis. "Margaret Mackenzie of Delvine, Perthshire: A Biographical Note on an Eighteenth-Century Weather Observer." *Journal of Meteorology* (U.K.) 19 (1994): 79–86.

———. "The Weather Diary of Margaret Mackenzie of Delvine (Perthshire): 1780–1805." *Scottish Geographical Magazine* 110 (1994): 177–82.

Wheeler, Roxann. *The Complexion of Race: Categories of Difference in Eighteenth-Century British Culture*. Philadelphia: University of Pennsylvania Press, 2000.

Widdowson, J. D. A. "Form and Function in Traditional Explanations of Weather Phenomena." In *Folklore Studies in Honour of Herbert Halpert*, edited by Kenneth S. Goldstein and Neil V. Rosenberg, 353–76. St. Johns: Memorial University of Newfoundland, 1980.

Widmalm, Sven. "Accuracy, Rhetoric, and Technology: The Paris–Greenwich Triangulation, 1784–88." In Frängsmyr, Heilbron, and Rider, *Quantifying Spirit*, 179–206.

Wilcox, Donald J. *The Measure of Times Past: Pre-Newtonian Chronologies and the Rhetoric of Relative Time*. Chicago: University of Chicago Press, 1987.

Williams, Raymond. *Culture and Society, 1780–1950*. Harmondsworth, Middlesex: Penguin Books, 1963.

Willis-Bund, J. W., and H. A. Doubleday, eds. *The Victoria History of the County of Worcester*. 5 vols. London: A. Constable and Co., 1901–26.

Willmoth, Frances. "John Flamsteed's Letter concerning the Natural Causes of Earthquakes." *Annals of Science* 44 (1987): 23–70.

Wilson, Kathleen. *The Island Race: Englishness, Empire and Gender in the Eighteenth Century*. London: Routledge, 2003.

Wilson, Stephen. *The Magical Universe: Everyday Ritual and Magic in Pre-Modern Europe*. London: Hambledon and London, 2000.

Wokler, Robert. "Anthropology and Conjectural History in the Enlightenment." In Fox, Porter, and Wokler, *Inventing Human Science*, 31–52.

Wolf, A. *A History of Science, Technology and Philosophy in the 18th Century*. Edited by Douglas McKie. 2nd ed. 2 vols. London: George Allen and Unwin, 1962.

Wood, Paul B. *The Aberdeen Enlightenment: The Arts Curriculum in the Eighteenth Century*. Aberdeen: Aberdeen University Press, 1993.

Woolf, D. R. "The 'Common Voice': History, Folklore, and Oral Tradition in Early Modern England." *Past and Present*, no. 120 (1988): 26–52.

Worden, Blair. "Providence and Politics in Cromwellian England." *Past and Present*, no. 109 (1985): 55–99.

Zealley, P. R. "A Florid Weather Diary, Worcestershire, 1703." *The Meteorological Magazine* 59 (1924): 91–92.

———. "A Florid Weather Diary." *Weather* 25 (1970): 555.